Broadcast and Two-Way Radio

Operators Permit Handbook

by

Edward M. Noll

A Revision of
Third-Class Radiotelephone License Handbook
by Edward M. Noll

Howard W. Sams & Co., Inc.

4300 WEST 62ND ST. INDIANAPOLIS, INDIANA 46268 USA

Contents

CHAPTER 1

RADIO OPERATORS 7
License Considerations—Essential Provisions for Radio Operators—Radiotelephone Distress Procedure

CHAPTER 2

TWO-WAY RADIO SERVICES 24
Maritime Radio Services—Frequencies and Range—Important Rules—Marine Radiotelephones—Aviation Radio Services—Public Safety Radio Services—Industrial Radio Services—Land Transportation Radio Services—Citizens Band Radio Service—General Radio Service

CHAPTER 3

RADIO BROADCAST OPERATION 72
Technical Considerations—The Operator's Responsibilities—Transmitter Metering—Operating Power—Station Plans—WFIL - WUSL—WGSA - WIOV— Monitoring Summary—Remote Control—State Requirements and Operator Duties—Log Requirements — Malfunctions—Summary —Additional Rules and Procedures

CHAPTER 4

BROADCAST STUDY GUIDE AND SELF-EXAMINATION . . . 111
Multiple-Choice Test—Answers to Multiple-Choice Test—
Study Questions—Self-Test—Answers to Self-Test

CHAPTER 5

THIRD-CLASS RADIOTELEPHONE LICENSE 127
Element I–Basic Law—Element II–Basic Operating Practice
—Elements I and II Self-Test—Answers to Self-Test

APPENDIX A

EXTRACTS FROM THE GENEVA, 1959, TREATY 145

APPENDIX B

EXTRACTS FROM THE COMMUNICATIONS ACT OF 1934,
AS AMENDED 147

APPENDIX C

EXTRACTS FROM THE FCC RULES AND REGULATIONS,
PARTS 1, 2, 13, 17, 81 AND 83 150

APPENDIX D

CITIZENS BAND (CB) RADIO SERVICE 165

INDEX 175

Preface

This handbook serves as a practical study guide for the aspiring radio operator, as well as a ready reference for those working in the field. The laws, rules, regulations, and accepted operating procedures, as they apply to radio operators in the various radio services are given in this volume. Information needed by the restricted radiotelephone permit holder as well as that needed by the third-class radiotelephone operator is included. This handbook can also serve as a guide for those anticipating a career in radio operations.

Since a knowledge of the internal operation, critical adjustments, and maintenance of radio transmitters and other electronic gear is not required to obtain these lower-grade licenses, this type of information has been omitted. These licenses are concerned only with the persons who operate the equipment and carry on radio communications.

Chapter 1 covers the various grades of operator licenses and explains how these licenses are obtained. In addition, many helpful operating rules and procedures are given in this chapter. The nonbroadcast services—Maritime, Aviation, Public Safety, Industrial, Land Transportation, and Citizens Band— are discussed in Chapter 2.

The additional information needed in the operation of broadcast stations is given in Chapter 3. The functions that can be performed by the lower-grade license holders, including the restricted radiotelephone operator permit, are presented in this chapter. Details are given about transmitter monitoring, operating procedures, log-keeping, and other information that will give you a better technical understanding of station operations. Although the operator may only be interested in announcing and other broadcast studio activities, proper operating procedures must be known if he must monitor the operation of the broadcast transmitter as a part of his activities.

Chapter 4 is a question-and-answer presentation that will help the operator reinforce his knowledge of operating procedures and techniques. Although some new information is given in Chapter 4, most of the coverage is a comprehensive review of the information given in Chapter 3.

Chapter 5 is for the operator who wishes to advance another step to the third-class radiotelephone license. This license permits the operator to assume additional responsibilities in the Maritime and Aviation services. It requires that the prospective operator pass Elements I and II of the FCC license examinations. Element I is concerned with basic law while Element II covers basic operating practices. The test for the third-class license is based on these two elements. At the conclusion of the text there is a simple self-test that can be taken to check out your knowledge of Elements I and II. This test is patterned after the FCC style of multiple-choice examinations.

A knowledge of radio law, including specific FCC Rules and Regulations, is essential to pass the examination and be a good operator. Important excerpts from these Rules and Regulations are given in Appendices A, B, and C. These rules should be studied by all operators.

Operators in the Citizens Band Radio Service now number in the millions. Considerable coverage of Citizens band operation and equipment is given in Chapter 2. Complete rules and regulations for the Citizens Band Radio Service are given in Appendix D. Every licensee in the Citizens Band Radio Service must have a copy of these rules according to FCC regulations. In the Citizens Band (CB) Radio Service a station license is required but not an operator's license. Nevertheless the operator is responsible for proper operation of CB equipment and Appendix D will give the details on this information.

All information needed to secure an FCC license up to, but not including, the second-class license is given in this handbook. Information is presented on how to obtain the appropriate permits and licenses. Additional information and hints of value to the operator have been included. You will learn how to be a proper and capable operator. This is a must book for anyone contemplating a career in radio operating or dispatching.

EDWARD M. NOLL

CHAPTER 1

Radio Operators

Radio operators number in the millions in this era of two-way radio growth. Radio operators are radio operators for a variety of reasons. If you pilot a small boat or fly a small plane, your reason for being a radio operator involves the safety and convenience offered by your two-way radio equipment. Such stations are licensed in the Maritime or Aviation Radio Services, respectively.

Perhaps you are a radio operator because two-way radio expedites your business procedures, as in the Industrial, Citizens, or other land radio services. Your radio operating reason may be a desire for personal-convenience communications (Citizens Band Radio Service), or a hobby activity (Amateur Radio Service). Your association with radio operations may be as a public service in one of the Public Safety Radio Services.

Some radio operators need not be licensed; others require an FCC license of a particular class. Whether or not you are a licensed operator, you are responsible for the proper operation of the station with which you are associated. It is your duty to know the appropriate laws, rules and regulations, permissible communications, and established procedures for the particular radio service with which you are concerned.

In 1979 a new FCC ruling permits the operation of fm and most am broadcast stations by the holder of a *restricted radiotelephone operator permit*. No license examination is required for this permit. Nevertheless it is advisable that you know the appropriate rules and regulations. Some basic technical knowledge is helpful too. Certain restrictions apply and the station must be suitably equipped for this manner of operation. Refer to Chapter 3.

LICENSE CONSIDERATIONS

This handbook is concerned with the nonlicensed operator and the lower-grade FCC licenses, up to, but not including, the second-class radio license. These lower-class licenses are concerned only with the operation of the equipment for carrying on communications. They do not require a knowledge of the technical aspects of two-way radio equipment.

Although the restricted license or permit can be obtained without examination, some technical background is often helpful and perhaps mandatory for advancement in the particular radio service.

Radio transmitters require a *station license*. The operator of the transmitter and station does, or does not, require an operator license depending on the radio service rendered. For example, in the Citizens Band Radio Service a station license is required, but the operator is not licensed. However, the operators of any radio transmitter are responsible to the licensee of the station, and operate with his authorization. Every operator, whether required to be licensed or not, must obey the rules governing the station he operates.

In the Aviation and Maritime service, both *station* and *operator* licenses are required. If you have a station license for a small boat or small plane, you also are required to have a restricted radio-telephone operator permit, as a minimum *operator license*. No examination is required for this license. In other mobile and maritime radio services, higher-grade licenses are required. These licenses involve taking an FCC examination to prove that the applicant possesses the required knowledge.

It is necessary to pass Elements I and II to obtain a *third-class operator permit*. This grade of license is required for those radio services in which it is imperative that the operator know the appropriate rules, regulations, and operating procedures. Such an examination is taken at a district office of the Federal Communications Commission. (For more information regarding license applications and examinations, the reader may refer to Chapter 5 of this book.)

To obtain a restricted radiotelephone operator permit, it is only necessary to fill out FCC Form 753. This form can be obtained from any district office of the FCC. The addresses and phone numbers of the FCC field offices are given in Chart 1-1. A sample of form 753 is shown in Fig. 1-1. This application is mailed to the Federal Communications Commission, Gettysburg, Pennsylvania 17325.

Specific operator requirements and authority as they appear in the FCC Rules and Regulations, Part 13, are as follows:

(f) *Radiotelephone third-class operator permit:*

(1) Ability to transmit and receive spoken messages in English.

(2) Written examination elements: 1 and 2.

(h) *Restricted radiotelephone operator permit:*
No oral or written examination is required for this permit. In lieu thereof, applicants will be required to certify in writing to a declaration which states that the applicant has need for the requested permit; can receive and transmit spoken messages in English; can keep at least a rough written log in English or in some other language in general use that can be readily translated into English; is familiar with the provisions of treaties, laws, and rules and regulations governing the authority granted under the requested permit; and understands that it is his responsibility to keep currently familiar with all such provisions.

License authorities are as follows:

(g) *Radiotelephone third-class operator permit.* Any station except:

(1) Stations transmitting television other than Instructional Television Fixed Service stations, or

(2) Stations transmitting telegraphy by any type of the Morse Code, or

(3) Any of the various classes of broadcast stations, or

(4) Class I-B coast stations at which the power is authorized to exceed 250 watts carrier power or 1000 watts peak envelope power, or

(5) Class II-B or Class III-B coast stations, other than those in Alaska, at which the power is authorized to exceed 250 watts carrier power or 1000 watts peak envelope power, or

(6) Ship stations or aircraft stations at which the installation is not used solely for telephony, or at which the power is more than 250 watts carrier power or 1000 watts peak envelope power:

Provided, that (1) such operator is prohibited from making any adjustments that may result in improper transmitter operation, and (2) the equipment is so designed that the stability of the frequencies of the transmitter is maintained by the transmitter itself within the limits of tolerance specified by

FCC FORM 753
SEPTEMBER 1975
PART I

FEDERAL COMMUNICATIONS COMMISSION
P. O. BOX 1050
GETTYSBURG, PA. 17325

APPROVED BY GAO
B-180227 (R0222)

APPLICATION FOR RESTRICTED RADIOTELEPHONE OPERATOR PERMIT BY DECLARATION

A. USE TYPEWRITER OR PRINT IN INK. Signatures must be handwritten. Be sure to complete all items including 7, 8, 9, and 10.
B. Enclose fee with application. See Fee Schedule. (FCC Form 76 - K). DO NOT SEND CASH. Make check or money order payable to FEDERAL COMMUNICATIONS COMMISSION.
C. No oral or written examination is required. Applicant must be at least 14 years of age.
D. Submit application - Parts I and II - to FCC, P. O. Box 1050, Gettysburg, Pa. 17325
E. ALIENS should submit FCC Form 755 instead of this form.

DO NOT WRITE IN THIS BLOCK

(1) REASON FOR APPLICATION

☐ NEW PERMIT

☐ NAME CHANGE
(Attach Present Permit)

☐ REPLACE PRESENT PERMIT
DUE TO ITS CONDITION
(Attach Present Permit)

☐ ORIGINAL PERMIT IS LOST OR DESTROYED. IF FOUND I WILL RETURN IT TO FCC. A REASONABLE SEARCH HAS BEEN MADE FOR THE PERMIT

☐ OTHER
(Specify)

(2) NAME (Last)   (First)   Middle Initial

PERMANENT ADDRESS   (No. & Street)

(City)   (State)   (ZIP Code)

IBM Z26441

APPLICANT'S CERTIFICATION

	MONTH	DAY	YEAR	
③ DATE OF BIRTH·				

④ ARE YOU A CITIZEN OF THE U. S. ?   YES ☐ NO ☐

⑤ Have you been convicted in the last ten years of any crime for which the penality imposed was a fine of $500 or more or sentenced to imprisonment of more than one year?   YES ☐ NO ☐

(IF YES, FURNISH DETAILS FOR EACH CONVICTION GIVING DATE. NATURE OF CRIME, COURT IN WHICH CONVICTED, NATURE OF SENTENCE AND WHERE SENTENCE WAS SERVED.)

⑥ Do you have any physical defects such as a speech impediment, acute deafness, or any other defect which will impair or handicap you in properly using the permit for which you are applying?   YES ☐ NO ☐

(IF "YES", ATTACH DETAILS.)

IBM Z26440

I certify that I am the above named applicant; that the facts stated in the foregoing application and all exhibits attached thereto, are true of my own knowledge; that I have need for the permit herein applied for; that I can transmit and receive spoken messages in English; that I can keep at least a rough written station log in English, or in some other language in general use that can be readily translated into English; that I am familiar with the provisions of treaties, laws, rules and regulations governing the authority granted under the permit herein applied for; that I understand that it is my responsibility to keep myself currently familiar with all such provisions, that I will preserve the secrecy of radio communications as required by law and that I will faithfully adhere to any requirements of law at all times, that this obligation is taken freely, without mental reservation or purpose of evasion; and that I will well and faithfully discharge the duties of the office obtained through my employment under this Permit if granted.

⑦ _______________________________   _______________
(Signature)   (Date)

WILLFULL FALSE STATEMENTS MADE ON THIS FORM ARE PUNISHABLE BY FINE AND IMPRISONMENT. U. S. CODE, TITLE 18, SECTION 1001

Chart 1-1. FCC Field Offices

Alaska
Anchorage
U. S. Post Office Bldg.,
Room G63
4th & G Street, P.O. Box 644
Anchorage, AK 99510
Phone (907) 265-5201

California
Long Beach
3711 Long Beach Blvd.,
Suite 501
Long Beach, CA 90807
Phone (213) 426-4451

San Diego
Fox Theatre Bldg.
1245 Seventh Ave.
San Diego, CA 92101
Phone (714) 293-5478

San Francisco
323A Customhouse
555 Battery Street
San Francisco, CA 94111
Phone (415) 556-7701

Colorado
Denver
Suite 2925, The Executive Tower
1405 Curtis St.
Denver, CO 80202
Phone (303) 837-5137

District of Columbia
Washington
1919 M Street N.W., Room 411
Washington, DC 20554
Phone (202) 632-8834

Florida
Miami
919 Federal Bldg.
51 S.W. First Ave.
Miami, FL 33130
Phone (305) 350-5541

Tampa
Barnett Bank Bldg.
1000 Ashley St.
Tampa FL 33602
Phone (813) 228-2872

Georgia
Atlanta
Room 440, Massell Bldg.
1365 Peachtree St. N.E.
Atlanta, GA 30309
Phone (404) 881-3084

Savannah
238 Federal Office Bldg. &
Courthouse
125 Bull St., P.O. Box 8004
Savannah, GA 31402
Phone (912) 232-4321, Ext 320

Hawaii
Honolulu
7304 Prince Jonah Kuhio
Kalanianaole Bldg.
300 Ala Moana Blvd.
Honolulu, HI 96813
Phone (808) 546-5640

Illinois
Chicago
3935 Federal Bldg.
230 S. Dearborn St.
Chicago, IL 60604
Phone (312) 353-0195

Louisiana
New Orleans
829 F. Edward Hebert
Federal Bldg.
600 South St.
New Orleans, LA 70130
Phone (504) 589-2094

Maryland
Baltimore
819 George M. Fallon
Federal Bldg.
31 Hopkins Plaza
Baltimore, MD 21201
Phone (301) 962-2728

Massachusetts
Boston
1600 Customhouse
165 State St.
Boston, MA 02109
Phone (617) 223-6609

Michigan

Detroit
1054 Federal Bldg. & U.S.
  Courthouse
231 W. Lafayette St.
Detroit, MI 48226
Phone (313) 226-6078

Minnesota

St. Paul
691 Federal Bldg.
316 N. Robert St.
St. Paul, MN 55101
Phone (612) 725-7810

Missouri

Kansas City
1703 Federal Bldg.
601 E. 12th St.
Kansas City, MO 64106
Phone (816) 374-6155

New York

Buffalo
1307 Federal Bldg.
111 W. Huron St.
  at Delaware Ave.
Buffalo, NY 14202
Phone (716) 842-3216

New York
201 Varick St.
New York, NY 10014
Phone (212) 620-3437

Oregon

Portland
1782 Federal Office Bldg.
1220 S.W. 3rd Ave.
Portland, OR 97204
Phone (503) 221-3908

Pennsylvania

Philadelphia
11425 James A. Byrne Federal
  Courthouse
601 Market St.
Philadelphia, PA 19106
Phone (215) 597-4411

Puerto Rico

Hato Rey (San Juan)
Room 747 Federal Bldg. &
  Courthouse
Avenida Carlos Chardon
Hato Rey, PR 00918
Phones (809) 753-4567 or
  753-4008

Texas

Beaumont
Room 323 Federal Bldg.
300 Willow St.
Beaumont, TX 77701
Phone (713) 838-0271, Ext 317

Dallas
Room 13E7 Earle Cabell
  Federal Bldg.
1100 Commerce St.
Dallas, TX 75242
Phone (214) 749-1719

Houston
5636 Federal Bldg.
515 Rusk Ave.
Houston, TX 77002
Phone (713) 226-5624

Virginia

Norfolk
Military Circle
870 N. Military Highway
Norfolk, VA 23502
Phone (804) 441-6472

Washington

Seattle
3256 Federal Bldg.
915 Second Ave.
Seattle, WA 98174
Phone (206) 442-7653

the station license, and none of the operations necessary to be performed during the course of normal rendition of the service of the station may cause off-frequency operation or result in any unauthorized radiation, and (3) any needed adjustments of the transmitter that may affect the proper operation of the station are regularly made by or under the immediate supervision and responsibility of a person holding a first- or second-class commercial radio operator license, either radiotelephone or radiotelegraph as may be appropriate for the class of station involved . . . , who shall be responsible for the proper functioning of the station equipment and (4) in the case of ship radiotelephone or aircraft radiotelephone stations when the power in the antenna of the unmodulated carrier wave is authorized to exceed 100 watts, any needed adjustments of the transmitter that may affect the proper operation of the station are made only by or under the immediate supervision and responsibility of an operator holding a first- or second-class radiotelegraph license, who shall be responsible for the proper functioning of the station equipment.

(h) *Restricted radiotelephone operator permit.* Any station except:

(1) Stations transmitting television, or

(2) Stations transmitting telegraphy by any type of the Morse Code, or

(3) Any of the various classes of broadcast stations other than fm translator and booster stations, or

(4) Ship stations licensed to use telephony at which the power is more than 100 watts carrier power or 400 watts peak envelope power, or

(5) Radio stations provided on board vessels for safety purposes pursuant to statute or treaty, or

(6) Coast stations, other than those in Alaska, while employing a frequency below 30 MHz, or

(7) Coast stations at which the power is authorized to exceed 250 watts, carrier power or 1000 watts peak envelope power;

(8) At a ship radar station the holder of this class of license may not supervise or be responsible for the performance of any adjustments or tests during or coincident with the installation, servicing, or maintenance of the radar equipment while it is radiating energy: *Provided,* That nothing in this subparagraph shall be construed to prevent any person holding such a license from making replacements of fuses or of receiving type tubes: *Provided,* That, with respect to any station which the holder of this class of license may operate, such operator is

prohibited from making any adjustments that may result in improper transmitter operation, and the equipment is so designed that the stability of the frequencies of the transmitter is maintained by the transmitter itself within the limits of tolerance specified by the station license, and none of the operations necessary to be performed during the course of normal rendition of the service of the station may cause off-frequency operation or result in any unauthorized radiation, and any needed adjustments of the transmitter that may affect the proper operation of the station are regularly made by or under the immediate supervision and responsibility of a person holding a first- or second-class commercial radio operator license, either radiotelephone or radiotelegraph, who shall be responsible for the proper functioning of the station equipment.

§ 13.62 Special privileges

In addition to the operating authority granted under § 13.61, the following special privileges are granted holders of commercial radio operator licenses:

(b) The holder of any class of radiotelephone operator's license, whose license authorizes him to operate a station while transmitting telephony, may operate the same station when transmitting on the same frequencies, any type of telegraphy under the following condition:

(1) When transmitting telegraphy by automatic means for identification, for testing, or for actuating an automatic selective signaling device, or

(2) When properly serving as a relay station and for that purpose retransmitting by automatic means, solely on frequencies above 50 MHz, the signals of a radiotelegraph station, or

(3) When transmitting telegraphy as an incidental part of a program intended to be received by the general public, either directly or through the intermediary of a relay station or stations.

(c) The holder of any class of commercial operator license may operate AM, FM, or educational FM broadcast stations, except AM stations using directional antenna systems which are required by the station authorizations to maintain ratios of the currents in the elements of the systems within a tolerance which is less than five percent or relative phases within tolerances which are less than three degrees, under the following conditions:

(1) That adjustments of transmitting equipment by such operators, except when under the immediate supervision of a radiotelephone first-class operator (radiotelephone second-class operator for educational FM stations with transmitter output power of 1000 watts or less), and except as provided in paragraph (d) of this section, shall be limited to the following:

(i) Those necessary to turn the transmitter on and off;

(ii) Those necessary to compensate for voltage fluctuations in the primary power supply;

(iii) Those necessary to maintain modulation levels of the transmitter within prescribed limits;

(iv) Those necessary to effect routine changes in operating power which are required by the station authorization;

(v) Those necessary to change between nondirectional and directional or between differing radiation patterns, provided that such changes require only activation of switches and do not involve the manual tuning of the transmitter's final amplifier or antenna phasor equipment. The switching equipment shall be so arranged that the failure of any relay in the directional antenna system to activate properly will cause the emissions of the station to terminate.

(2) The emissions of the station shall be terminated immediately whenever the transmitting system is observed operating beyond the upper and lower limiting values of parameters required to be observed and logged or in any manner inconsistent with the rules or the station authorization, and the above adjustments are ineffective in correcting the condition of improper operation, and a first-class radiotelephone operator is not present.

(3) The special operating authority granted in this section with respect to broadcast stations is subject to the condition that there shall be in employment at the station in accordance with Part 73 of this chapter one or more first-class radiotelephone operators authorized to make or supervise all adjustments, whose primary duty shall be to effect and ensure the proper functioning of the transmitting system. In the case of a a noncommercial educational FM broadcast station with authorized transmitter output of 1000 watts or less, a second-class radiotelephone licensed operator may be employed in lieu of a first-class licensed operator.

(d) When an emergency action condition is declared, a person holding any class of radio operator license or permit, who is authorized thereunder to perform limited operation of a standard broadcast station, may make any adjustments necessary to effect operation in the emergency broadcast system in accordance with the station's National Defense Emergency Authorization: *Provided,* That the station's responsible first-class radiotelephone operator(s) (or second-class radiotelephone operator(s), for noncommercial educational FM broadcast stations with transmitter output power not exceeding 1 kilowatt) shall have previously instructed such person in the adjustments to the transmitter which are necessary to accomplish operation in the Emergency Broadcast System.

ESSENTIAL PROVISIONS FOR RADIO OPERATORS

All radio operators, licensed or not, should know and are responsible for knowing the following laws and regulations.

Station License

A radio station, other than one belonging to and operated by the Federal Government, shall not be operated unless it is properly licensed by the Federal Communications Commission and the station license is posted or kept available as specified by the rules governing the particular service and/or class of station.

Operator Licenses or Permits

Except as may be provided otherwise by the rules governing a particular service and/or class of station, a radio station required to be licensed by the Commission shall be operated only by a properly licensed radio operator who has his license or verification card or his permit in his possession or posted in accordance with the Commission's rules governing the particular service in which he is employed. If the license or permit has been sent in to the Commission for replacement, duplicate, etc., a copy of the application for such replacement, duplicate, etc., shall be exhibited in lieu of the license.

The holder of a restricted radiotelephone operator permit as issued to a United States citizen may operate any station in the fixed and mobile services while using radiotelephony, except:

1. Coast stations, other than in Alaska, while using a frequency below 30 MHz; or
2. Ship stations licensed to use telephony at which the power is more than 100 watts carrier power or 400 watts peak envelope power.
3. Radio stations provided on board vessels for safety purposes pursuant to statute or treaty.

Coast stations in the above categories which this grade of operator may operate are limited to those at which the power is not authorized to exceed 250 watts carrier or 1000 watts peak envelope power. The transmitting equipment of any station must be so designed that the stability of the operating frequencies is maintained by the transmitter itself within the limits of tolerance specified by the Commission; adjustments to the radio transmitter, which may cause off-frequency operation or result in improper transmitter operation, shall be made only by, or in the presence of, a person holding a first- or second-class operator license, either radiotelephone or radiotelegraph, who shall be responsible for the proper operation of the equipment.

Nature of Communications

Only such communications as are authorized by the rules governing the radio station operated may be transmitted. False calls, false or fraudulent distress signals, superfluous and unidentified communications, and obscene and profane language are specifically prohibited.

Priority of Communications

Distress calls and messages shall have absolute priority over all other communications. Distress calls may be made without regard to interference to other stations, with due consideration, however, being given to any other distress calls or messages which may be transmitted at the same time. Routine operation shall not be resumed until the distress signals and messages have been cleared.

The order of priority for communications in the mobile service shall be as follows:

1. Distress calls, distress messages, and distress traffic.
2. Communications preceded by the urgency signal.
3. Communications preceded by the safety signal.
4. Communications relating to radio direction-finding.
5. Communications relating to the navigation and safe movement of aircraft.
6. Communications relating to the navigation, movements, and needs of ships, and weather observation messages destined for an official meteorological service.
7. Government radiotelegrams: Priorite Nations.
8. Government communications for which priority has been requested.
9. Service communications relating to the working of the

radiocommunication service or to communications previously exchanged.

10. Government communications other than those shown in 7 and 8 above, and all other communications. (Art. 37.)

Secrecy of Radiocommunications

The contents of a radiocommunication shall not be divulged to any person or party other than to whom it is addressed, except as specifically provided in section 605 of the Communications Act.

Identification of Communications

When not required to identify itself by some other provisions of the Rules and Regulations, every radio station shall identify itself by its regularly designated call signal or other approved method at the time of each transmission, and as frequently as is practicable during tests or during an exchange of long communications.

Prevention of Interference

Inasmuch as most radio transmission in the mobile services is conducted on radio channels which are shared by many stations, as on a "party line," certain precautions must be observed to avoid unnecessary congestion and interference.

In order to avoid interference with communications in progress, an operator shall listen on the frequencies on which he intends to receive for a period sufficient to ascertain that he will be able to hear the station he is calling and that his transmission will not cause harmful interference. He shall not attempt to call if interference with established communications is likely to result.

Attempts to establish communication beyond the normal range of installed equipment usually result in unnecessary occupation of the calling frequency. Except in emergencies, such calling should be avoided.

In order to eliminate the need for undue repetition of communcations, voice transmission should be made with maximum articulation. It is well to remember that speech is generally rendered almost unintelligible by speaking too close to the microphone, and it is often lost in extraneous noise when the microphone is held at too great a distance.

Radio Log

Radio logs are required to be kept in certain radio services. These logs must be kept by a person having actual knowledge of the facts to be entered who shall also sign the log as prescribed by the Commission. Logs shall be made available on request by authorized Commission representatives. No log or portion thereof shall be erased, obliterated, or wilfully destroyed within the period of retention required by the Rules and Regulations. Any necessary correction may be made only by the person originating the entry; he shall strike out the erroneous portion, initial the correction made, and indicate the date of correction.

Notice of Violations

Any licensee who appears to have violated any provision of the Communications Act of 1934, as amended, or of the Rules and Regulations of the Federal Communications Commission, shall be served with a notice calling the facts to his attention and requesting a statement concerning the matter. Within 10 days from receipt of such notice, or such period as may be specified, the licensee shall send a written answer to the Commission field office, or to the monitoring station originating the official notice. If an answer cannot be sent or an acknowledgment made within such 10-day period by reason of illness or other unavoidable circumstances, acknowledgment and answer shall be made at the earliest practicable date, shall be complete in itself, and shall not be abbreviated by reference to other communications or answers to other notices. If the notice of violation relates to lack of attention to, or improper operation of, the transmitter, the name and license number of the operator shall be given.

Penalties

The general penalty for violation of the Communications Act (first offense) consists of a fine of not more than $10,000 or imprisonment for a term of not more than one year, or both.

The penalty for violation of the Commission's regulations or the international radio regulations consists of a fine of not more than $500 for each and every day during which such offense occurs.

The Commission has authority, as public convenience, interest, or necessity requires, to suspend the license (permit) of any operator on proof sufficient to satisfy the Commission that the operator:

(a) Has violated any provision of any act, treaty, or convention binding on the United States, which the Commission is authorized to administer, or any regulation made by the Commission under any such act, treaty, or convention; or

(b) Has failed to carry out a lawful order of the master or person lawfully in charge of the ship or aircraft on which he is employed; or

(c) Has wilfully damaged or permitted radio apparatus or installations to be damaged; or

(d) Has transmitted superfluous radio communications or signals or communications containing profane or obscene words, language, or meaning, or has knowingly:

    1. Transmitted false or deceptive signals or communications; or

    2. Transmitted a call signal or letter which has not been assigned by proper authority to the station he is operating; or

(e) Wilfully or maliciously interfered with any other radio communications or signals; or

(f) Obtained or attempted to obtain, or has assisted another to obtain or attempt to obtain, an operator's license by fraudulent means.

Distress Procedure

The international radiotelephone distress signal consists of the spoken expression MAYDAY. This signal shall be used to announce that the ship, aircraft, or other vehicle that sends the distress signal is threatened by serious and imminent danger and requests immediate assistance. The distress signal shall be followed by the distress messages containing the identity of the station in distress, its position, the nature of the distress, and the nature of the assistance requested.

The international radiotelephone urgency signal consists of the word PAN, spoken three times, and is to be used when the calling station has a very urgent message to transmit concerning the safety of the ship, aircraft, or other vehicle, or concerning the safety of some person on board or sighted from on board.

The international radiotelephone safety signal consists of the word SECURITY spoken three times, and announces that the station is about to transmit a message concerning the safety of navigation, or giving important meteorological warnings.

The distress call *sent* by radiotelephony comprises:

1. The distress signal MAYDAY spoken three times;
2. The words THIS IS, followed by the identification of the mobile station in distress, the whole repeated three times.

The distress call should be followed as soon as possible by the distress message, which comprises:

1. The distress signal MAYDAY;
2. The name of the ship, aircraft, or vehicle in distress;
3. Particulars of its position, the nature of the distress, and the kind of assistance desired;
4. State number of adults and children aboard and conditions of any injured.
5. Any other information which might facilitate the rescue.

As a general rule, a ship shall signal its position in latitude and longitude, using figures for the degrees and minutes, together with one of the words NORTH or SOUTH and one of the words EAST or WEST.

As a general rule, and if time permits, an aircraft shall transmit in its distress message the following information: estimated position and time of the estimate; heading in degrees (state whether magnetic or true); indicated air speed; altitude; type of aircraft; nature of distress, type of assistance desired, and any other information which might facilitate the rescue (including the intention of the person in command, such as forced alighting on the sea or crash landing).

After the transmission by radiotelephony of its distress message, the mobile station may be requested to transmit suitable signals followed by its call sign or other identification, to permit direction-finding stations to determine its position. This request may be repeated at frequent intervals if necessary. In radiotelephony, the signal is made by holding the transmitter on the air for the specified period of time, without speaking into the microphone other than to identify the station.

Immediately before a crash landing or forced landing (on land or sea) of an aircraft, as well as before total abandonment of a ship or an aircraft, the radio apparatus should be set for continuous emission, if considered necessary and circumstances permit.

RADIOTELEPHONE DISTRESS PROCEDURE

A radiotelephone distress procedure as recommended by the Radio Technical Commission for Marine Services is given in Chart 1-2.

DISTRESS COMMUNICATIONS FORM

Instructions: Complete this form now (except for items 6 through 9) and post near your radiotelephone.

Speak SLOWLY — CLEARLY — CALMLY

1. Make sure your radiotelephone is on.

2. Select either *VHF Channel 16* (156.8 MHz) or *2182* kHz.

3. Press microphone button and say: "MAYDAY — MAYDAY — MAYDAY."

4. Say: "THIS IS ______________, ______________, ______________,
   your boat name   your boat name   your boat name
   ______________."
   your call letters

5. Say: "MAYDAY: ______________."
   your boat name

6. TELL WHERE YOU ARE (What navigational aids or landmarks are near?).

7. STATE THE NATURE OF YOUR DISTRESS.

8. GIVE NUMBER OF ADULTS AND CHILDREN ABOARD, AND CONDITIONS OF ANY INJURED.

9. ESTIMATE PRESENT SEAWORTHINESS OF YOUR BOAT.

10. BRIEFLY DESCRIBE YOUR BOAT:

    ______________; ______________ FEET;
    State Registration No.   Length

    ______FEET; ______; ______HULL; ______TRIM;
    Draft   Type   Color   Color

    ______MASTS; ______POWER; ______.
    Number   Type; Hosepower   Construction Material

    Anything else you think will help rescuers to find you.

11. Say: "I WILL BE LISTENING ON *CHANNEL 16/2182*."
    (Cross out channel no. or frequency that does not apply)

12. End Message by saying: "THIS IS ______________ OVER,"
    your boat name and call sign

13. Release microphone button and listen: Someone should answer. IF THEY DO NOT, REPEAT CALL, BEGINNING AT ITEM 3. If there is still no answer, switch to another channel and begin again.

Two-Way Radio Services

This chapter covers each of the major two-way radio services and categories with particular emphasis on operator license requirements and responsibility. Discussions of typical operations and installations are given. There is some coverage of station license considerations, plus operating frequency bands and important spot frequencies for the various radio services.

MARITIME RADIO SERVICES

The maritime industry was the first enthusiastic and worldwide user of two-way radio. Early in this century radio stations were installed on sea-going vessels. Military, government, and private land stations were set up to maintain contact with these radio-equipped vessels. Obviously, the two major divisions of the maritime service are stations on land and stations on shipboard. There are, of course, a number of subdivisions of both land and shipboard stations.

Two-way radio is compulsory on most vessels. There must be an efficient radio installation in operating condition and in charge of, and operated by, a qualified operator or operators on any ship of the United States (other than a cargo ship of less than 300 gross tons) to be navigated in the open seas outside of a harbor or port. No such vessel may leave or attempt to leave any harbor or port of the United States for a voyage on the open seas unless it is so equipped.

Furthermore, any vessel, regardless of size, that is transporting more than six passengers for hire and is navigated in the open seas or any tidewater within the jurisdiction of the United States adjacent or contiguous to the open seas must be equipped with an acceptable radio installation. The FCC may exempt from the provision of this part any vessel or class of vessel where the route or condition of the voyage or other condition or circumstances are such as to render a radio installation unreasonable, unnecessary, or ineffective.

Most vessels on which a radio installation is compulsory must include a main radiotelegraph installation and, in most cases, an emergency or reserve radiotelegraph installation. For cargo ships of less than 1600 gross tons, a radiotelephone may be installed in lieu of a radiotelegraph installation.

Three major categories of coast stations are: public, limited, and marine-utility stations. The public and limited stations are further subdivided into three additional categories. Classes I, II, and III. A public coast station is one that is said to be open to public correspondence. Public correspondence itself refers to any telecommunications which the offices and stations must, by reason of their being at the disposal of the public, accept for transmission. When suitable rates are filed, it is possible to charge for such correspondence. A limited coast station is one that is not open to public correspondence but serves mainly the operational and business needs of ships.

Be it public or limited, a Class-I coast station provides a mobile radio service to ships at sea, including such service over distances up to several thousand miles. Class-I coast stations are the only ones that have frequency assignments below 150 kHz or between 5000 and 25,000 kHz. A Class-II coast station provides a maritime service primarily of a regional character. Frequency assignments are not made below 150 kHz or between 5000 and 25,000 kHz. The Class-III coast station provides a maritime service of a local nature and does not operate on any frequency below 25 MHz. The marine-utility station is one that is readily portable for use as a limited coast station at unspecified points ashore within a designated local area.

Two main categories of ship stations are also public and limited. A limited ship station is not open to public correspondence and must confine itself to the operational and business needs of shipping. Public ship stations are also further classified according to their hours of service for telegraphy in a public correspondence service. These categories are based on whether they provide continuous service of public correspondence or a limited service.

The equipment aboard those vessels on which a radio installation is compulsory must meet specific requirements. The installations are inspected and are issued safety certificates attesting to their compliance with the radio requirements of the Safety Convention.

Operator Requirements

In general the classes of operator licenses required for the compulsory shipboard radio installations are above the license classes which are the major concern of this handbook. For the classes of ship and coast stations covered in the previous paragraphs, it is apparent that most operators must have a radiotelegraph license. Except for certain low-power coast stations, the grade of license must be at least a second-class radiotelegraph grade.

Some vessels require only a compulsory radiotelephone installation and a minimum grade of third-class radiotelephone operator license is required. Details on this grade of license are given in Chapter 5. When the power of the ship's station exceeds 250 watts carrier or 1500 watts peak envelope power the operator shall have a minimum second-class radiotelephone operator license. Certain of the harbor land-mobile stations which are associated with a larger coast station do not require a licensed operator. However, in this case the communications are under the control of the properly licensed operator of the main station.

One might conclude from the previous discussion that lower license grades have no place in the maritime radio service. However, there are many types of vessels excluded from the compulsory radio installations. These are the teeming numbers of small commercial and pleasure boats, sailboats, certain yachts, etc. Two-way radio installations on these boats require a station license and an operator's license. The station license application is made on FCC Form 502; the operator license application is made on Form 753. The operator license is a restricted radiotelephone operator's permit and it is obtained without an examination. The fact that the license does not require an examination does not excuse the holder of a restricted radiotelephone operator's permit from knowing the laws, rules and regulations, and operating procedures associated with the particular radio services.

The primary objective of two-way radio in the Maritime Radio Service is safety of life and property. Thus, all operators are responsible for knowing the distress procedures covered in Chapter 1.

FREQUENCIES AND RANGE

Marine radios for recreational use operate in the vhf spectrum (156–162 MHz) or medium frequency (mf) spectra between 2–3 MHz. The fm-vhf marine radios are popular for small-boat applications. For longer transmission range the single-sideband mf radios are employed. If you incorporate an mf radio you must also incorporate an fm-vhf unit.

Range of transmission on the vhf marine band (156–162 MHz) is approximately 10–30 miles ship-to-shore and 10–15 miles ship-to-ship. Daytime range of transmission on the 2–3-MHz medium-frequency marine band is 50–150 miles depending upon a number of variables such as available power, noise, and interference. When a greater range of operation is desired, there are additional medium-frequency marine bands: 4-, 6-, 8-, 12-, and 16-MHz allocations. A competent dealer can provide additional information for operating on these bands.

IMPORTANT RULES

Let us consider some of the rules in detail. The radiotelephone transmitter of a ship station must be operated by a licensed operator. The licensed operator may permit others to speak over the microphone if he starts, supervises, and ends the operation, makes the necessary log entries, and gives the necessary identification.

The license usually held by radio operators aboard small vessels not required to carry a radio installation for safety purposes is the restricted radio operators permit. This is a lifetime permit. However, it does not authorize transmitter adjustments that may affect the proper operation of the station. Any needed adjustments must be made only by the holder of a first- or second-class radiotelegraph or radiotelephone license. It is not necessary to post the restricted radiotelephone operator permit if it is kept on the operator's person. However, other classes of licenses must be conspicuously posted at the principal location at which the station is operated.

The frequency of 2182 kHz is the calling and distress frequency. Ship radiotelephone stations in the 2–3-MHz band must maintain an efficient listening watch on this frequency while the station is open and not transmitting on other frequencies. All shipboard transmitters in this band must be capable of transmitting on this frequency. There are also certain intership frequencies that may be employed. These frequencies are limited to use for safety and operational communications, and in the case of commercial transport vessels they may be used for business communications.

Whenever a marine radio is turned on, the receiver should be tuned to the distress frequency, 156.8 MHz or 2182 kHz, if the radio is not being used for communications. Watch times must be recorded in the radio log. A radio silent period on 2182 kHz must be maintained for three minutes immediately after the hour and for three minutes immediately after the half hour. Only messages concerning distress or urgency can be transmitted during these silent periods.

Coast Guard

Recreational boating operators use the frequencies of 2670 kHz and 156.1 MHz (channel 22A) for communicating with coast guard stations. Contact is first established with the Coast Guard by calling them via the appropriate calling frequency of 2182 kHz or 156.8 MHz (channel 16).

Operating Procedure

A number of definite operating procedures must be followed. Before transmitting, always listen on the channel that you intend to use so as to minimize interference. You must give your call sign whenever you call another vessel or coast station, and also when you finish the conversation. Make your calls short (not more than 30 seconds) and do not call the same station again for two minutes. If a call sign has not as yet been assigned, you may identify your station by announcement of the vessel name and the name of the licensee.

If you hear a radio conversation not intended for you, you cannot lawfully use the information in any way. Do not forget that safety is the primary reason for having shipboard radio. Distress and safety must have absolute priority. This is the reason for setting aside distress frequencies. You transmit on this frequency when you are in distress, and you maintain a watch on this frequency so that you might help another in distress.

Radio Log

It is necessary to keep a radio log. Each page of the log must be numbered, must have the name and call sign of the vessel, and must be signed by the operators. The start and end of the watch on 2182 kHz and 156.8 MHz must be recorded. All distress and alarm signals and related communications transmitted or intercepted and all urgency and safety signals and related communications transmitted shall be recorded in the log as completely as possible.

A record of all installations, and service or maintenance work performed, which may affect the proper operation of the station, must also be entered by the licensed operator doing the work, including his signature, address, class of license, serial number, and expiration date of his license. Use the 24-hour system in the radio log; that is, 8:45 am is written as 0845 and 1:00 pm becomes 1300.

Radio logs must be retained for at least one year—three years if they contain entries concerning distress or disaster. If at any time you receive an official notice of violation from the FCC, you must reply to it within ten days.

Transmitters

Each radiotelephone transmitter used in a ship station must be type-accepted under Part 83 of the Commission's Rules and Regulations. Except for transmitting equipment required to comply with Title III, Part II of the Communications Act, no application for modification of license is required for the deletion, addition, or replacement of radiotelephone and radar transmitters which operate in the frequency bands specfied on the license. The additional or replacement transmitters must be type-accepted or type-approved, as appropriate.

Transmitters for the 2–3-MHz band must be capable of single-sideband emission. Each station must also be equipped for transmission in the 156–158-MHz band.

2182 Kilohertz

This is the calling and distress frequency for ship radiotelephone stations in the 2–3-MHz band, and these stations must maintain an efficient listening watch on 2182 kHz while the station is open and not transmitting on other frequencies. All ship stations in this band must be capable of transmitting on 2182 kHz.

Recommended Channels for Ship Stations Using Channels in the 2–3-MHz Band

Ship radiotelephone stations in the 2–3-MHz band, if used for other than safety communications, shall be capable of transmitting on at least two working frequencies. There are eight intership frequencies in the band, as shown in Table 2-1.

Use of these intership frequencies is limited to safety and operational communications, except that commercial transport vessels may use them also for business communications.

Table 2-1. Intership Frequencies

Frequency (kHz)	Geographic area
2003	Great Lakes only.
2082.5	All areas.
2142	Pacific coast area south of latitude 42 degrees north, on a day only basis.
2203	Gulf of Mexico.
2638	All areas.
2670	All areas.
2738	All areas except the Great Lakes and the Gulf of Mexico.
2830	Gulf of Mexico only.

156.8 Megahertz, Channel 16

This is the calling and distress frequency for ship radiotelephone stations in the 156–162-MHz band. These stations must maintain a watch on 156.8 MHz and be capable of transmission on 156.8 MHz, 156.3 MHz, and one or more working frequencies. Channel 6 is used for intership safety communications. Both of these channels are mandatory.

Recommended Channels for Ship Radiotelephone Stations Using Channels in the 156–158-Megahertz Band

Ship radiotelephone stations in the 156–158-MHz band must be capable of transmitting on 156.8 MHz, 156.3 MHz, and at least one more working frequency. Table 2-2 has been prepared as a guide to assist you in deciding what channels to install in your vhf radio, and to determine which channels to use in a particular situation. The table covers both commercial and recreational usage for radios with 4-, 6-, 8-, 12-, 16-, or 24-channel capability. The final selection will be determined by comparing this table with the facilities available in your area of interest.

MARINE RADIOTELEPHONES

Small marine transceivers are operated in the 2–3-MHz and 156–158-MHz bands. These compact ship-to-shore radiotelephones can be installed conveniently on small commercial and pleasure boats. The unit shown in Fig. 2-1 is a completely synthesized vhf-fm model capable of operation on 55 transmit channels and 96 receive channels. In addition there are four weather receive channels. The unit can be mounted out of the weather on a bulkhead, overhead, or on a table top.

Transmit power is the 25-watts maximum permitted on the vhf band. In addition the power can be switched down to 1

Table 2-2. Channel Designations

Channel Designators	TYPE OF COMMUNICATIONS Points of communications	Commercial vessels Channel capability						Recreational vessels Channel capability					
		4	6	8	12	16	24	4	6	8	12	16	24
		No. of recommended channels of each group Select channels used in area of operation											
16 (mandatory)	DISTRESS, SAFETY & CALLING Intership & ship to coast	1	1	1	1	1	1	1	1	1	1	1	1
06 (mandatory)	INTERSHIP SAFETY Intership	1	1	1	1	1	1	1	1	1	1	1	1
65, 66, 12, 73, 14, 74, 20, 05	PORT OPERATIONS Intership & ship to coast		1	2	2	3	6		1	1	2	3	7
13	NAVIGATIONAL Intership & ship to coast		1	1	1	1	1				1	1	1
15	ENVIRONMENTAL Ship receive only				1	2	2			1	1	2	2
17	STATE CONTROL Ship to coast				1	1	1				1	1	1
07, 09, 10, 11, 18, 19, 79, 80	COMMERCIAL Intership & ship to coast	1	1	1	2	2	4						
67, 08, 77, 88	COMMERCIAL Intership			1	1	1	3						
68, 09	NONCOMMERCIAL Intership & ship to coast							1	1	1	1	1	2
69, 71, 78	NONCOMMERCIAL Ship to coast								1	1	1	1	3
70, 72	NONCOMMERCIAL Intership										1	2	2
24, 84, 25, 85, 26, 86, 27, 87, 28	PUBLIC CORRESPONDENCE Ship to public coast	1	1	1	2	4	5	1	1	2	2	3	5

watt for minimum interference during in-harbor operation. Proper operating frequency is obtained by dialing up the correct channel number with the synthesizer switches. There is a quick-select Channel-16 switch that enables instant use of the emergency channel. A red indicator light comes on for Channel 16 operation. One other switch position is made available for quick-select operation. A green light comes on when this channel is activated.

A digital synthesized marine radiotelephone is shown in Fig. 2-2. Direct dialing of the channel is possible using a keyboard-entry microprocessor control. A digital readout indicates the operating channel. Transmitter frequency range extends between 155.5 and 158.975 MHz; receiver range, between 155.5 and 162.6 MHz. The number of transmit channels is 55; receiver channels, 104.

In addition you can set up automatic scanning of 20 keyboard-selected channels. Also a priority watch can be established on any channel by keyboard selection. For example, this

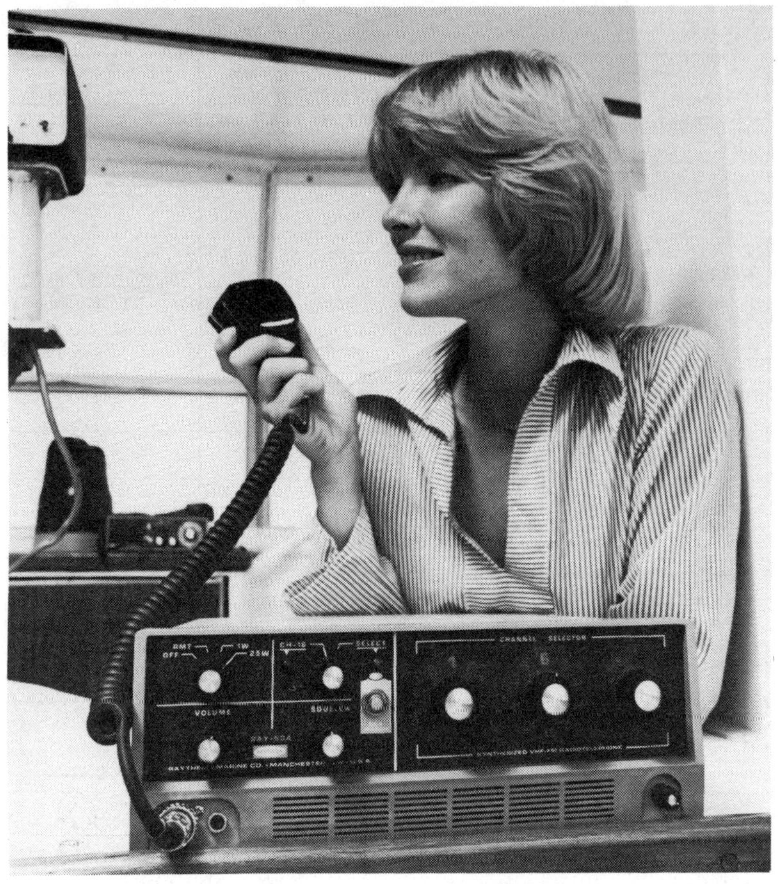

Courtesy Raytheon Marine Co.

Fig. 2-1. A vhf-fm 25-watt marine radiotelephone.

latter feature could be used to provide a watch on Channel 16. Transmitter high power is 25 watts; low power, 1 watt.

A hand-held model such as that shown in Fig. 2-3 can be used on small boats with no power supply, as an emergency unit, or for short-range transmissions. It includes its own battery supply. It is capable of 6-channel operation using crystals. Usually Channels 6 and 16 are supplied with the unit when purchased. Crystals for other channels can be purchased. The unit weighs only 2.2 lbs (1 kg) including its optional rechargeable nickel-cadmium power pack. This model can also be operated for a limited length of time on eight AA penlight batteries.

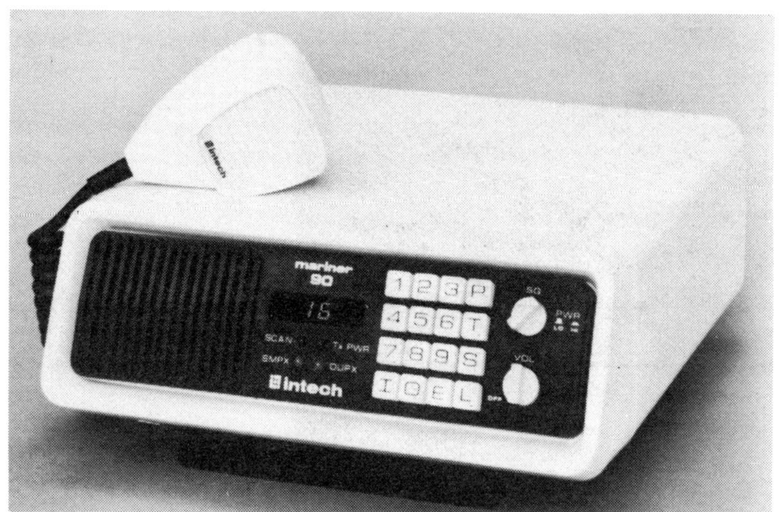

Courtesy Intech Inc.

Fig. 2-2. Synthesized scanning marine radiotelephone.

Short-range marine communications now take place more frequently on the vhf rather than the mf bands. In fact, a vhf installation is compulsory even though mf band equipment is to be installed on the boat. The mf band has changed over completely to single-sideband transmission rather than the old double-sideband, amplitude-modulation technique which required a greater operating bandwidth. Single-sideband operation requires only one-half of the emission bandwidth needed by standard amplitude modulation.

A 65-watt single-sideband 2–4-MHz radiotelephone is shown in Fig. 2-4. This 65-watt transmitter power compares favorably with the older 250-watt double-sideband transmitter. It includes eight crystal-controlled channels including the 2182-kHz international emergency frequency.

A reputable dealer can help you select crystals for the channels most useful in your area. For example, if the radiotelephone were to be used on the East Coast, crystals could be installed on frequencies that would give you East Coast coverage from Maine to South Carolina. In addition there are Coast Guard, ship-to-ship, as well as key public coast channels.

An mf installation is appropriate for those boats that go outside of the range of vhf-fm gear such as sport fishing boats, small commercial craft and coastal cruisers.

A receiver of the type shown in Fig. 2-5 is popular on small boats. It is a portable direction finder which operates on the

Fig. 2-3. Hand-held marine transceiver.

Courtesy Apelco Marine Electronics

marine radio-beacon, marine radiotelephone, and radio broadcast bands. This receiver in conjunction with a chart can be used to take compass bearings on selected stations. By obtaining a fix on two or more stations you can determine the position of your vessel. It is a simple and inexpensive navigation aid. There are over 200 Coast Guard marine radiobeacons that broadcast distinct signals that can be picked up by such a direction-finding unit. Fixes can be made on nearby coastal broadcast stations too.

A sensitive ferrite loop antenna is used for direction finding in conjunction with a vertical-rod sense antenna. This model can be supplied by a 12-volt power source or eight 1.5-volt D cells.

Courtesy Raytheon Marine Co.

Fig. 2-4. Single-sideband marine radiotelephone.

AVIATION RADIO SERVICES

The Aviation Radio Services set aside portions of the spectrum for radiocommunication and radionavigation facilities for aircraft operators, aeronautical enterprises, and organizations that require radio-transmitting facilities for safety purposes and other necessities. The radio stations are allocated on the basis of a number of categories—airborne-aeronautical advisory, aeronautical multicom, aeronautical enroute, aeronautical metropolitan, flight test, flying school, airdrome, aeronautical utility, aeronautical search and rescue mobile, aeronautical fixed, operational, radionavigational land, and civil air patrol.

Most of these stations require a licensed operator. Again, insofar as installation and maintenance are concerned, a higher grade license is needed. If radiotelegraphy is employed, the operator must have at least a third-class radiotelegraph license. Certain stations require a second-class radiotelegraph license.

In general, communications in the aviation services shall be restricted to safe, expeditious, and economical operation of aircraft, and the protection of life and property in the air. Some of the Aviation Radio Services, namely aeronautical public service, aeronautical advisory, aeronautical multicom stations, and civil air patrol land and mobile stations may conduct additional communications in accordance with the particular service of which they are a part.

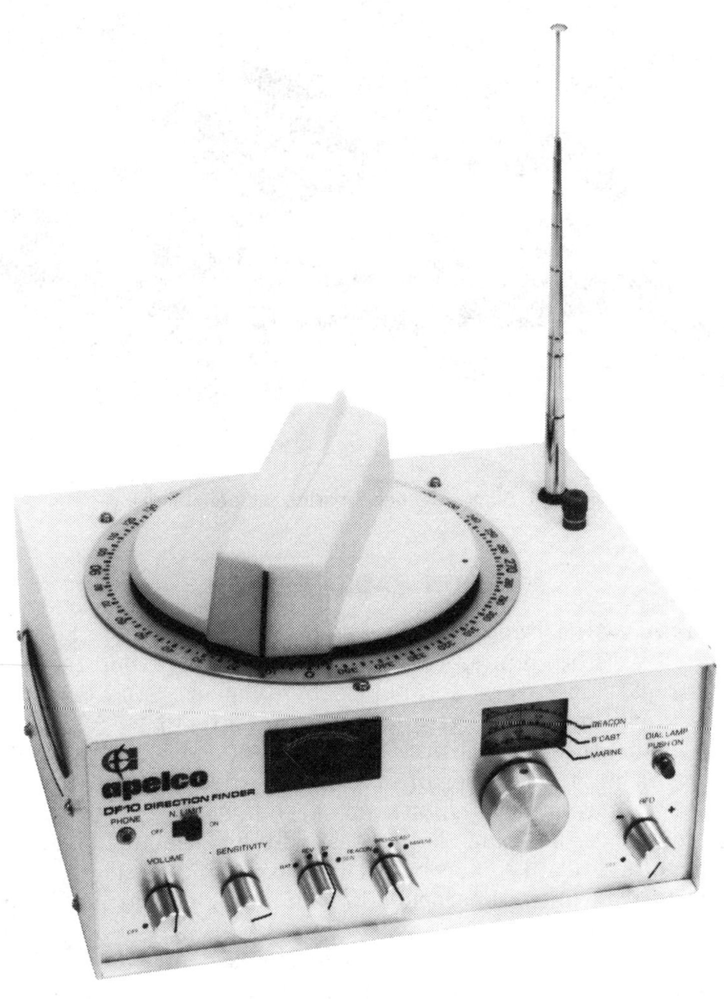

Courtesy Apelco Marine Electronics

Fig. 2-5. Direction-finding receiver.

As in other radio services, each transmitter must be licensed. Except for certain land mobile units used at airdromes and for other aeronautical applications, each station must be operated by a licensed operator. Quite often only a *restricted radiotelephone operator permit* is needed. Although no examination is required in obtaining a permit, as mentioned in Chapter 1, the operator is responsible for knowing the laws and FCC rules and regulations that apply to the particular radio service with which he is associated.

In summary, application must be made for both station and operator licenses for airborne and ground aeronautical radio stations. FCC Forms 404, 406 and 753 are used.

Operator Requirements

Practically all airborne and aeronautical ground radio stations require that the operators be licensed. In most cases the license is only a restricted radiotelephone permit. As covered previously, no examination is required, and the license is issued for life. For example, such a license is required if you fly a private aircraft that employs two-way radio. Such an operator license is also required for the operator of a two-way radio used at a small private airfield.

Again, although the operator holds only a restricted permit, he must be familiar with the appropriate operating procedures and rules that apply to the particular radio station. Installation, maintenance, and any adjustments that influence the radiation from the transmitter must, of course, be made by an operator with a second-class radiotelephone license or higher. If radiotelegraphy or other form of coded transmission is used, the operator must have an appropriate radiotelegraph license or permit.

For higher-powered stations and more complex equipment using directional antennas and other facilities, higher grade licenses are often required. Usually for transmitters with authorized powers in excess of 250 watts carrier or 1000 watts peak envelope power, the minimum grade license is a second-class type.

Station Categories

There are a variety of airborne and ground aeronautical stations. The four classifications of airborne stations are: air carrier, private, flight test and flying school, and aeronautical public service. An air-carrier aircraft station is one aboard an aircraft engaged in, or essential to, the transportation of passengers or cargo for hire. A private aircraft station is one aboard an aircraft not operated as an air carrier.

An exception to the allocation of air carrier or private aircraft stations involves the weight of the aircraft. If it is less than 12,500 pounds, it may be considered, at the option of the applicant, as a private aircraft even though it is actually engaged in air-carrier operations.

Obviously, station licenses are also granted to aircraft used at test facilities or in flying schools. Licenses can also be obtained for aircraft used for the handling of public correspondence in the same manner as such services are available to ships in the Maritime Radio Service. These stations can handle messages for hire. Facilities must be made available for the use of all persons and without discrimination; such stations shall intercommunicate with any other stations similarly licensed when necessary for the handling of traffic.

Aircraft radio stations using radiotelephony, with the exceptions noted in § 87.133 of the FCC rules, shall be operated by persons holding any class of commercial radio operator license or permit. All transmitter adjustments or tests during, or coincident with, the installation, servicing, or maintenance of a radio station that may affect the proper operation of such station, shall be made by or under the immediate supervision or responsibility of a person holding a first- or second-class commercial radio operator license, either radiotelephone or radiotelegraph, who shall be responsible for the proper functioning of the station equipment.

There is even a greater variety of aeronautical ground stations. Airdrome control stations provide communications limited to the necessities of safe and expeditious operation of aircraft using the airdrome facilities or operating within the airdrome control area. Such stations are required to monitor the following frequencies during hours of operation—121.5 MHz and 3023.5 kHz (Alaska only).

In association with the airdrome there are also certain land mobile stations called aeronautical utility and aeronautical search and rescue mobile stations. Such stations can be installed in fuel trucks, airdrome repair trucks, emergency vehicles, etc.

Throughout the United States there are numerous radionavigation land stations that guide and give instructions to aircraft enroute. Most of these stations are operated by the Federal Aviation Agency. These stations along with the airdrome radio facilities provide radionavigation coverage that extends from destination to destination, both on and off the civil air routes.

For this fundamental purpose there are certain other station classifications operated by the FAA, or through private ownership if the applicant justifies the need for the facility and the government is not prepared to render this service. These stations are aeronautical enroute, aeronautical metropolitan, aeronautical fixed, and operational stations.

Small airports and landing areas can be allocated aeronautical-advisory and aeronautical-multicom station licenses. When operated at a landing area not served by an airdrome control station, communications must be limited to the necessities of safe and expeditious operation of private aircraft, pertaining to the condition of runways, types of fuel available, wind conditions, weather information, dispatching, or other necessary information. On a secondary basis, information concerning ground transportation, food, or lodging can also be handled.

The same information can be communicated via an aeronautical advisory station in an area served by an airdrome control station, with the exception that information concerned with runway conditions, fuel, and weather are handled by the airdrome control station.

A multicom station can be used to direct activities associated with such aeronautical activities as fighting forest fires, aerial advertising, parachute jumping, etc.

Ground base facilities can also be provided for flying schools, flight-test facilities, and Civil Air Patrol.

All of the various aeronautical ground stations just covered must be operated by FCC-licensed operators, with the exception of the FAA stations operated by government employees. Also, certain land mobile stations do not require an operator; usually such vehicles are associated with an airdrome control station or other aeronautical ground station which is under the control of a licensed operator.

Aeronautical Radiocommunication and Radionavigation

There are many electronic aids to aircraft navigation; they can be considered in the four general categories of communication, navigation, traffic control, and landing. Most of the modern-day aeronautical radio activity occurs in the frequency spectrum between 108 and 136 MHz. Radionavigation uses the spectrum between 108 and 118 MHz; air traffic control, 118 to 136 MHz. Spotted throughout this spectrum are frequencies assigned to both aircraft and aeronautical ground stations. For example, in flying a private aircraft you will find the frequencies of the various ground stations given on navigation maps and/or charts. These aeronautical ground stations monitor certain aircraft frequencies and you can quickly establish contact enroute by setting your aviation radio to an appropriate frequency.

Many aviation radio units have dual-reception facilities. Thus, it is possible to receive a radio navigation signal continuously at the same time that a two-way radio contact is being made with an aeronautical ground station. Such a unit is often referred to as a one and one-half communicator because it has a single transmitter and two receivers.

Most modern flying is done via the vhf Omnidirectional Range stations. These are called VOR or OMNI stations. In the vhf frequency spectrum there is largely static-free reception, and the bending and false beams of the older low-frequency radio range stations are not present. A reliable directional pattern can be produced at these frequencies. A complex revolving antenna pattern that uses electronic switching generates a rotating beam that has a directional accuracy of 1° throughout the entire 360° of rotation.

A reasonably complete Narco installation in a Piper Cherokee Arrow is shown in Fig. 2-6. The two vertical rows of basic equipment begin immediately to the right of center on the aircraft control panel. The radio and navigation switcher is located at the top of the first panel row, followed by two communicators. Each communicator contains a transmitter and receiver that operate on the frequency indicated by its digital

Fig. 2-6. Aircraft radio installation.

readout. Thus, each communicator can be used as a simplex transceiver operating on the same frequency. However, duplex operation is possible by setting one communicator to the desired transmit frequency and the second communicator to a designated receive frequency. In this manner of operation the transmit and receive frequencies differ.

The navigation receiver is located immediately below the two communicators. Associated meters indicate bearing and right-left positioning of aircraft with respect to the VOR range station being received by the navigation receiver.

A transponder is the last unit of this row. When a transponder is activated it sends out a coded signal. At a ground station, aircraft can be seen and identified on a radar screen. A very strong signal is received because the transponder is activated whenever the aircraft is searched for by the strong beam from the ground-based radar station. Furthermore the return signal is a coded one and this information is also displayed. Therefore, the ground station is not only able to determine the distance and direction to the plane but also is able to make a positive identification of the particular plane because of its returned transponder signal.

At the top of the second row is a radio direction finder. This receiver can be tuned to the low-frequency range station. Its associated meter (top and immediately to the left of the first panel of equipment) shows the relative bearing of the range station with respect to the aircraft heading.

The bottom of the second panel is a distance-measuring (DME) unit. In operation it sends out an interrogating signal to a special range facility called a Vortac station. The Vortac station, in turn, sends back a signal to the aircraft unit. By utilizing the time of travel of the interrogation signal and the return signal, it is possible to determine the distance to the Vortac station. The actual range in nautical miles is indicated by the meter on the left side of the instrument. A ground-speed indicator, the meter at the lower left of the panel, operates in conjunction with the distance-measuring unit. It reads ground speed in flying toward or away from the Vortac station.

Emergency and Distress

121.5 MHz is a universal simplex clear channel for use by aircraft in distress or condition of emergency. It will not be assigned to aircraft unless other frequencies are assigned and available for normal communications needs. The channel is available, as follows:

1. For emergency communications when circumstances beyond the control of the pilot prevent communication between the aircraft and ground stations on other regularly assigned channels.
2. For emergency direction-finding purposes.
3. For establishing air-to-ground contact by aircraft in distress, emergency, or when lost.
4. In connection with search and rescue operations, to provide a common channel for aircraft (either civil or

military) not equipped to transmit on 123.1 MHz. This includes communications between aircraft, and between aircraft and ground stations. Stations having the capability should change to 123.1 MHz as soon as practicable.
5. To provide a common frequency for survival communications and for survival radio beacons (emission A2).
6. For air/ground communications between aircraft and ocean station vessels for safety purposes when service on other vhf channels is not available.

The frequency 243 MHz is available to survival craft stations which are also equipped to transmit on 121.5 MHz.

Equipment Tests

Aircraft stations are authorized to make routine tests when required for the proper maintenance of the station, provided that precautions are taken to avoid interference with any other station. A call from an aircraft in flight on the frequency 121.5 MHz for the purpose of making an unannounced or unanticipated test of the alertness of a ground station is not permitted.

Station Operation

Private and air carrier aircraft radio stations are generally limited to communications relating to the necessities of safe aircraft operation. Aeronautical public service stations may be authorized for use aboard aircraft to provide a means of conducting public correspondence. The licensee of a radio station is responsible at all times for the proper operation of his station. Thousands of other aircraft stations use the same frequencies that are assigned to your station. Be brief; transmit only essential messages. Your calls will receive quicker response, repeats will be fewer, and a general improvement in aircraft safey communications will result if the following precautions are taken:

1. Shorten or eliminate test calls while on the ramp or in flight.
2. Be sure the channel is clear before transmitting.
3. Tune your receiver to the correct receiving channel before transmitting.

Station Identification

Aircraft stations frequently cause confusion by failure to properly identify themselves when calling or working by radiotelephone. Aircraft radio stations are normally identified by use of the FAA registration number. After the first communication of each series, the last two characters of the registration number may be used if the practice is initiated by the ground station operator.

PUBLIC SAFETY RADIO SERVICES

The Public Safety Radio Services provide a service of radiocommunication essential to the discharge of either nonfederal governmental functions or the alleviation of an emergency endangering life or property. FCC licensed as well as nonlicensed operators and dispatchers are required for these services.

Specific radio station assignments are made under the categories of local government, police, fire, special emergency, highway, forestry-conservation, and state guard. It is to be noted that many of these radio services are closely related. Hence, they are suited to the setting up of radio control centers. For maximum benefit and minimum interference and confusion, there should be close cooperation in the selection of operating frequencies.

The radio operator in the Public Safety Service must be a good operator and develop a good understanding of police and emergency procedures. The Public Safety Radio Services are rather closely knit in a manner similar to the individual services that come under both the marine and aviation radio services. For example, a police radio operator or dispatcher should also be familiar with fire and special-emergency radio procedures. The Industrial and Land-Transportation radio categories encompass more divergent fields of interest than the Public Safety Radio Services.

In summary, the public service radio operator is much more closely linked with the general public in comparison to the get-a-job-done objectives of the Land Transportation and Industrial Radio Services, or the safe-journey objectives of an aircraft or boat station in the aviation or marine radio services. The public safety radio operator and displatcher is a responsible public servant.

Operator Requirements

Public safety radio stations can be operated by unlicensed persons when so authorized by the station licensee. Such operators may be stationed at either control or dispatch points.

For each of the radio services a station license must be obtained. Application is made on FCC Form 400. The exact FCC regulations are as follows:

§ 89.163   Operator requirements

(a) *Operation during the course of normal rendition of service—radiotelephone.*

(1) The following classes of stations may be operated by an unlicensed person, if authorized to do so by station licensee:

(i) From a control point—a mobile, a base or fixed station.

(ii) From a dispatch point—a base or fixed station.

*　　*　　*　　*　　*　　*　　*　　*

(e) *Licensed operator required.* Notwithstanding any other provisions of this section, unless the transmitter is so designed that none of the operations necessary to be performed during the course of normal rendition of service may cause off-frequency operation or result in any unauthorized radiation, and unless the transmitter is so installed that all controls which may cause improper operation or radiation are not readily accessible to the person operating the transmitter, such transmitter shall be operated by a person holding a first- or second-class commercial radio operator license, either radiotelephone or radiotelegraph as may be appropriate for the type of emission being used, issued by the commission.

Frequency Advisory

The almost continuous activity of the Public Safety Radio Services requires wise conservation of frequency spectrum and effective operating procedures. Coded signals and special fill-in form transmissions are prevalent to conserve air time and expedite safety activities quickly and effectively. Frequency advisory committees are especially necessary to permit the wise allocation of frequencies and minimum interference among stations that share the same frequency or operate on nearby channels.

An operating-procedure manual has been published under the direction of a national organization. The Associated Public-Safety Communications Officers, Inc. (APCO). This organization acts in a frequency-advising capacity and has done much to consolidate and standardize law enforcement methods and operating procedures as they involve communications.

Radio-Control Centers

The need and growth of public safety radio operations has forced the development of large urban and county radio-control centers. The radio-control center of the police radio system of Bucks County, Pennsylvania, meets the requirements of an

Fig. 2-7. Police-dispatch control for Bucks County, Pennsylvania.

expanding suburbia (Fig. 2-7). It is an accelerated growth area that was a rural county only a few years ago. Public Safety Radio Services have been coordinated into a radio-control center, which includes police and fire radio control points (Fig. 2-8), ambulances, and paramedics. Civil defense and other special emergency radio services are housed at the

Fig. 2-8. Fire-dispatch control for Bucks County, Pennsylvania.

45

Fig. 2-9. Montgomery County, Maryland, emergency operating center.

Fig. 2-10. Pennsylvania State Police base station and emergency generator.

same location. Emergency power is available for the installation in case of power failure.

The township and municipalities are a part of the radio system that utilizes 13 vhf and 10 uhf transmitters. Interzone and coordination communication facilities are also available. A staff of approximately 75 people is required to operate the system which provides 24-hour service. Approximately 200,000 emergency calls are handled annually with a total of well over one million individual transmissions.

Such radio-control centers provide a high order of efficiency and effective coordination. The shared and cooperative use of frequencies reduces interference and speeds message handling. Coordination with other public-safety radio services reduces confusion and permits smooth handling of disaster activities.

A very unusual emergency radio-operating center has been constructed at Rockville in Montgomery County, Maryland (Fig. 2-9). It is a solid, well-built installation protected from adverse weather, heat, shock, nuclear radiation, etc. Facilities can house a staff of 75 persons for thirty days on stocked supplies. Power is made available by its own generating plant. Police, fire, and rescue radio systems are a part of the installation, including their respective dispatching rooms. The radio systems are in operation continually on a county-wide basis.

State police, because of the larger land area under their jurisdiction, must rely on two-way radio for rather long-

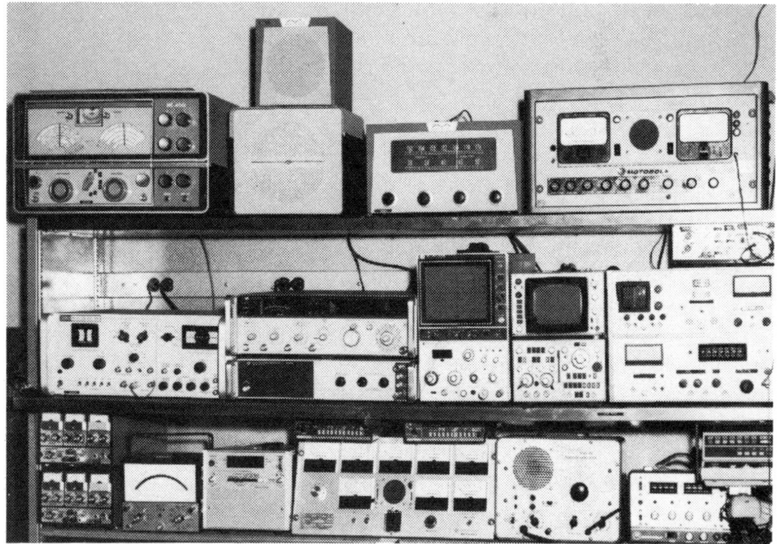

Fig. 2-11. Pennsylvania State Police two-way radio maintenance facility.

distance liaison as well as more localized communications. Base stations (Fig. 2-10) are strategically located around the state in a manner that can facilitate local and interzone communications. Some radio-relay facilities are also installed along turnpikes and express highways. Elaborate systems such as these require maintenance facilities. A representative test bench and array of test equipment are shown in Fig. 2-11.

INDUSTRIAL RADIO SERVICES

In the Industrial Radio Services part of the radio spectrum has been set aside for radiocommunication and control facilities for various industrial enterprises which, for safety purposes or other necessity, need radio-transmitting facilities in order to function efficiently. These radio services may not be used for a common-carrier service or to carry program material of any kind that will be used in any way in connection with radio broadcasting.

The radio stations are allocated on the basis of a number of categories—power, petroleum, forest, motion picture, relay press, special industrial, business, industrial radio location, manufacturers, and telephone-maintenance radio services. The four general classifications of stations are: mobile-base, fixed, mobile-relay, and fixed-relay stations. The mobile service, as in the radio services discussed previously, involves communications between mobile and base stations or among mobile stations. A fixed-station radio service is a service of radio-communications among specified fixed locations. A fixed relay station is established to receive radio signals directed to it from any source, and to retransmit them automatically for reception at one or more fixed locations. A mobile relay station is, in effect, a base station in the mobile service which is authorized primarily to retransmit automatically, on a mobile service frequency, communications originated by mobile stations.

The functional block diagram of Fig. 2-12 demonstrates the purpose of the four major industrial station types. Notice that the fixed relay station is a retransmission point located between two fixed stations. However, the mobile relay station is a permanent station that is used to provide a service of retransmission for a given mobile system where some of the mobile units must be operated at a further distance than the basic coverage area of the installation, or on the opposite side of an obstruction to radio propagation.

In the industrial radio services, permissible communications are those considered essential to the efficient conduct of that

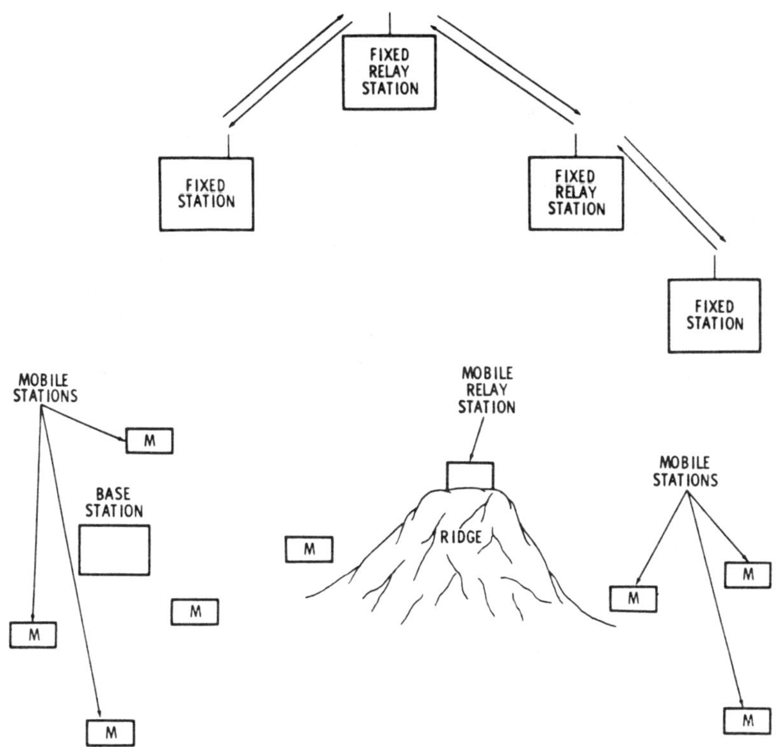

Fig. 2-12. Types of industrial radio stations.

portion of the enterprise for which the licensee is eligible to hold a station license, subject to the condition that harmful interference is not caused to safety communications of stations licensed in these services.

Control of a fixed relay station or a remotely located base station can be handled by a master controller (Fig. 2-13) positioned at a center of activity for a business or industrial complex. Using tone signalling over a telephone line as many as 12 control functions can be handled. The unit includes microphone, speaker, PTT bar, volume control, audio meter, digital clock, and selector switches for the function provided.

Permissible communications are also those related directly to the safety of life or the protection of property. In fact, a station licensee under this part may communicate with other stations without restrictions as to type, service, or licensee, when the communication to be transmitted involves safety of life or the protection of property.

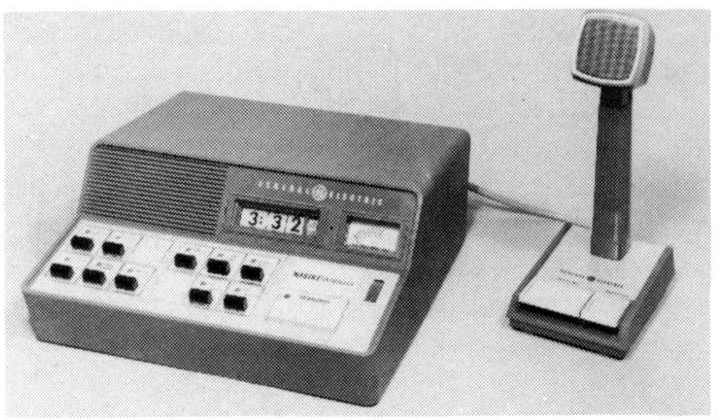

Courtesy General Electric Co.

Fig. 2-13. A master controller.

In the Industrial Radio Services, frequency assignments are allocated within the following bands: 1.6 to 6 MHz, 25 to 50 MHz, 152 to 174 MHz, and 450 to 460 MHz. For some of the services it is only necessary to make application using FCC Form 400. In other services, the form must be submitted together with evidence of frequency coordination. Frequency coordination, when required, must consider all stations operating on the requested frequency within 75 miles of the proposed station, and all stations operating on any adjacent frequency within 15 kHz of the requested frequency and within 35 miles.

Most stations are allocated for what is referred to as simplex operation. In simplex operation (Fig. 2-14), mobile and base stations operate on the same frequency. In a conversation only one station can transmit at a time. Usually this is referred to as push-to-talk operation, where a given mobile or base station

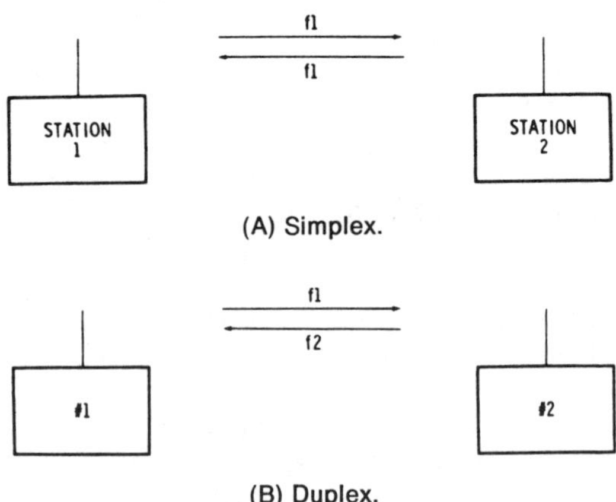

(A) Simplex.

(B) Duplex.

Fig. 2-14. Simplex and duplex operation.

switches between transmit and receive with the use of a microphone switch. In some of the Industrial Radio Services full duplex operation is permitted. In this mode of transmission more than one frequency is used and the radiocommunication can be telephonelike with both transmitters and receivers operating continuously during the communications.

Operator Requirements

During the course of normal rendition of service by radiotelephone, an unlicensed person, if authorized by the station licensee, may operate a base or fixed station from a dispatch point, or may operate a mobile, base, or fixed station from a control point. (If normal operation requires adjustments that could cause improper operation, or if the controls for such adjustments are readily accessible to the operator, then the operator must hold a first- or second-class commercial radio operator license. Also, the use of unlicensed operators does not relieve the station licensee of the responsibility for the proper operation of the station.) Only a person holding a commercial radiotelegraph license or permit (of any class) shall operate a station during the course of normal rendition of service when transmitting radiotelegraphy by any type of the Morse code.

Again, it is important to understand that all transmitter adjustments or tests during or coincident with the installation, servicing, or maintenance of a radio station, which may affect the proper operation of such station, shall be made by or under the immediate supervision and responsibility of a person holding a first- or second-class commercial radio license (must be radiotelegraph license for radiotelegraph stations transmitting by any type of Morse code).

The control point is the key location of a two-way radio installation. It is this point at which the person immediately responsible for the operation of the transmitter is located, and appropriate monitoring facilities are installed. This position must be under the control and supervision of the licensee.

The monitoring facility at the control point must include a carrier-operated device which will provide a visual indication when the transmitter is radiating or a pilot lamp or meter which will provide continual visual indication when the transmitter control circuits have been placed in a condition to produce radiation. At the control point the person responsible for the operation of the transmitter must be able to aurally monitor all transmissions originating at dispatch points. Facilities must be included which will permit the person responsible for the operation of the transmitter to disconnect dispatch-point circuits, and to turn the transmitter carrier on and off at will.

Station identification is also the control operator's responsibility. The assigned call letters should be transmitted at the end of each transmission or exchange of transmissions, or once every 15 minutes of the operating period, whichever the licensee may prefer. Mobile stations may use simple unit identifiers which must be kept on file in the station records of the associated base station.

Station Categories

Three of the categories of the Industrial Radio Services have to do with natural products; these are the power, petroleum, and forest-product services. Power-radio allocations are made to persons engaged in the generation, transmission, or distribution of electrical energy or to persons engaged in the distribution of manufactured or natural gas by means of a pipeline. The service also includes those engaged in the distribution of water or steam. It also applies to the activities associated with the collection, transmission, storage, or purification of water or the generation of steam preparatory to such distribution. The pertoleum radio allocations are made to those engaged in prospecting for, producing, collecting, refining, or transporting by means of pipeline, petroleum or petroleum products.

It is apparent that in these systems one can anticipate the use of a number of fixed radio stations, because of the great distances over which these products are transported by pipeline. Of course, base-mobile installations are essential at many locations, such as main distribution centers, prospecting sites, etc. Allocations are available for those persons engaged in tree logging, tree farming, or related woodland operations.

In a large industry a subsidiary corporation can be set up to handle radiocommunications on a nonprofit basis for the parent company or other subsidiaries of the parent corporation. It is also permissible to set up a separate nonprofit corporation or association to handle radiocommunications for persons engaged in one or more of the acceptable activities of the various radio services. Cooperative arrangements are also acceptable when made between two or more persons for the use of radio station facilities. All such persons must be eligible to hold a station license in one of the Industrial Radio Services. It is possible then to share the use of a base station which is licensed to one member of the group. Fixed stations can be operated in the same manner.

In the industrial radio category there are three rather closely related industrial and business services. These are: special industrial, manufacturers, and business radio services. In the manufacturers category the manufacturing activity should include the mechanical or chemical transformation of organic or inorganic substances into new products within establishments usually described as plants, factories, shipyards, or mills, and which employ in that process, power-driven machines and material-handling equipment. The radio service also includes those establishments engaged in assembling components of manufactured products in plants, factories, shipyards, or mills.

The manufacturers radio service does not apply to establishments primarily engaged in the wholesale or retail trade, or in service activities even though they fabricate any or all of the products or commodities handled. These activities are appropriate to the business radio service, which includes any person engaged in a commercial activity. The business radio service also applies to professional people; educational, philanthropic, or ecclesiastical institutions; and hospitals, clinics, and other medical associations.

Still another category is the special-industrial radio service. Allocations are set aside for persons regularly engaged in the operation of farms, ranches, or similar land installations, or for persons engaged in the operation of mines. Special industrial also applies to commercial business operations engaged in the construction of roads, bridges, sewers, pipelines, airfields, etc. Special-industrial allocations are also made to those engaged in specialized services essential to industrial operations or public health, such as soil conservation, seeding, fertilizing, spraying, etc. This service also extends to such varied activities as patrolling and repairing of gas and liquid transmission pipelines, water disposal systems, distribution systems of public utilities, cementing, logging, supplying of materials and services to large industrial organizations, delivery of ice or fuel, delivery and pouring of ready-mixed concrete or hot asphalt, etc.

It is apparent that almost any industrial activity is included somewhere in the list of permissible uses for the various industrial radio services. The business radio service has become increasingly popular because of the great help of two-way radiocommunications in professional and service activities. Many wholesale or retail businesses use two-way radio to great advantage. Modern equipment is efficient and very reliable. The complete transceiver shown in Fig. 2-15 is compact and can be mounted conveniently below the dashboard of the car. Battery-powered hand-held units, such as shown in Fig. 2-16, can be used by a person who must move about a large area on foot.

A more elaborate business radio system has been installed by Bux-Mont Communications of Hatfield, Pennsylvania. This installation was made for Glenn E. Garis, a builder and developer. A 100-watt low-band base station provides coverage south to Philadelphia and north to Allentown. There are five mobile stations. The General Electric transmitter is remotely controlled from two dispatch locations in the general offices. These are small compact units and include a telephone handset, as shown in Fig. 2-17. The trim base station installation by Bux-Mont Communications is shown in Fig. 2-18. Note that the sliding rack provides easy accessibility for service.

Courtesy Motorola, Inc.

Fig. 2-15. Business radiotelephone for mobile installation.

A typical mobile installation for the industrial radio services is shown in Fig. 2-19. Microphone, control head, and loudspeakers are mounted beneath the dashboard of the truck. The main transceiver is mounted at some other convenient position in the vehicle.

For most activities in the industrial radio services, the equipment can be operated by unlicensed persons. Or course, any such operation is only legal where it is authorized by the licensee who is responsible for the correct operation of the radio station. Just as you are responsible for knowing the

Fig. 2-16. A hand-held two-way radio.

Courtesy General Electric Co.

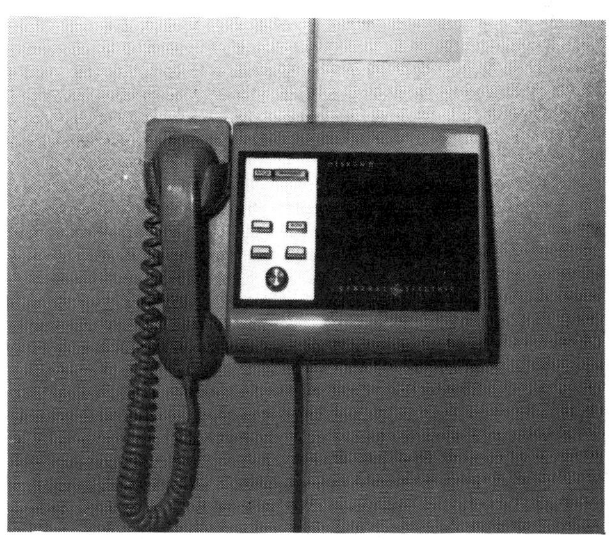

Fig. 2-17. Remote-control dispatch unit in office.

Fig. 2-18. Equipment rack at transmitter site.

law pertinent to the operation of an automobile, you are also responsible for knowing the law that governs the operation of a radio station that may be a part of the very same automobile.

Two other categories in the industrial radio services involve the news and entertainment media of the motion-picture industry and the press. Allocations in the motion-picture radio service are made to those engaged in the production or filming of motion pictures. Relay press allocations are made to persons engaged in the publication of a newspaper or any operation of an established press association.

There are frequency allocations for mobile telephone systems as well as telephone maintenance service. These frequencies are assigned to those engaged in rendering wire-line and radio communications service to the public for hire.

In Pennsylvania the progressive independent Denver and Ephrata Telephone and Telegraph Company takes full advan-

Fig. 2-19. Control head, microphone, and speaker for a mobile installation.

tage of these frequency allocations. Their modern installations provide mobile telephone and radio paging, as well as their own telephone maintenance services.

The base-station transmitter of the telephone maintenance installation operates in the 450-MHz band and its transmitter

Fig. 2-20. IMTS mobile telephone.

is unattended and located on a ridge not too far from the central office. This location provides a transmitter site with high average terrain relative to the service area of the telephone company. The transmitter is remotely controlled from the central office. Automatic switching and monitoring facilities are installed at the central office. There are 17 mobile units and three in-building dispatch points.

The same ridge site is the location of the high-band (150-MHz) transmitter of the mobile telephone system that provides service for mobile telephones, Fig. 2-20, which can be used locally and with the national mobile telephone service of the Bell system. Locally the system is completely automated and does not require the services of a mobile telephone operator. The automatic control panel is shown in Fig. 2-21. Details on the IMTS system follow in the next section.

The D&E Telephone Company also operates two completely automated paging systems. One is a tone-only facility and the other is a tone-voice combination. The automatic control panel for the tone paging system is shown in Fig. 2-22.

The paging receiver is a cigar-size unit that you can carry in the inside pocket of your suit jacket. If there is a message, you are alerted by a tone signal. With the tone-voice model,

Fig. 2-21. IMTS automatic control panel, D&E Telephone Co.

Fig. 2-22. Automatic control panel for tone-paging system, D&E Telephone Co.

you will receive actual voice messages after the alerting tone. Each unit is assigned a personal paging number by the telephone company and you can be dialed directly from any telephone.

One might at first think that the range of transmission would be very limited. The practical range of transmission is about a 15–20-mile radius. This range can be extended and better coverage obtained within the confined areas of large buildings by strategic location of more than one transmitter. The nearly 1000 D&E Telephone Company paging subscribers are served by one pair of transmitters located on the Ephrata ridge and another pair at a second site near the city of Lancaster. Two transmitters are needed at each site because the tone-only and the tone-voice systems operate on their own FCC-assigned frequencies.

Paging systems are automated and do not require the services of an operator. A breakdown alerts maintenance personnel automatically.

IMTS Telephone System

An effective mobile dial telephone system is in operation. This is referred to as the improved mobile telephone system (IMTS). For this service there are 11 channels allocated in the 150-MHz band. When such a telephone is installed in your car or other land vehicle, you will be able to dial any tele-

phone in your area without contacting a mobile service operator. Furthermore, when traveling into other cities you will be
able to dial into their radio system and make calls to any-
telephone in the area. It is anticipated that the system will
become nationwide and you will be able to dial any telephone
in this country from any location.

A basic plan of such a system is shown in Fig. 2-23. For
a large coverage area there is a centrally located group of

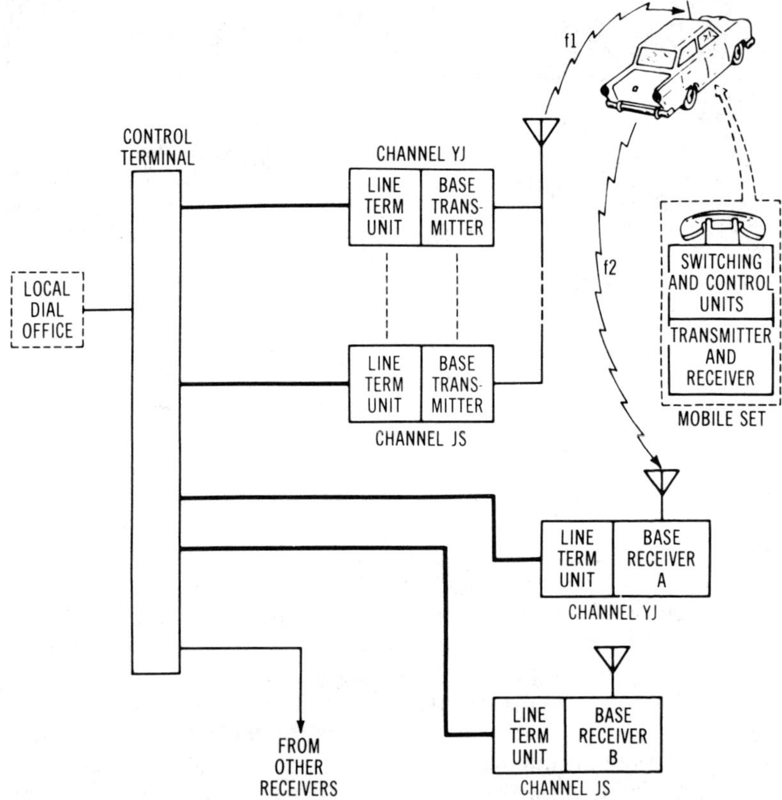

Fig. 2-23. IMTS system plan.

high-powered transmitters, one for each channel, and in any
given location up to a maximum of eight channels. Base receivers spotted conveniently around the area feed signals over
wire lines to the central office. The most favorable signal is
switched into the telephone system for use. In small-coverage
areas it is possible that a single low-powered base transmitter
and single receiver at the same site are adequate.

Each mobile unit consists of the transmitter-receiver combination and control unit, plus the channel-switching facilities and signaling circuits. Each unit is equipped to operate on at least all of the channels active in a phone area. For subscribers who are on the road, the mobile set may be equipped with additional channels up to a maximum of 11.

A special and very significant feature of the IMTS system is a feature referred to as multichannel access. A mobile unit automatically seeks an idle channel. Thus, when a mobile unit is activated, it will automatically seek an idle channel and establish a link from mobile to base station. The next call in either direction will then be completed over that channel. When this channel is occupied, all other mobiles in the area will automatically seek another idle base-station channel to establish still another connection. It is anticipated that each channel will be able to serve a maximum of 40 to 60 mobile units. The entire operation is automatic, and in most cases it is not necessary to call through a mobile service operator.

LAND TRANSPORTATION RADIO SERVICES

The Land Transportation Radio Services set aside parts of the radio spectrum to be employed for radiocommunication and radio-control facilities in certain land-transportation operations. Such facilities may not be used either to render a common-carrier communications service or to carry broadcast-program material.

Communications must be related directly to the safety of life or the protection of important property. Permissible communications may also involve the efficient operation of the transportation system described in the application and defined in the rules of eligibility for the particular service.

In general, both the station and operator requirements are the same as those applicable to the Industrial Radio Services.

Station Categories

There are four major categories in the Land Transportation Service; these are motor carrier, railroad, taxicab, and automobile emergency radio services. In the taxicab radio service, certain frequencies are set aside for base and mobile units and for mobile use only. Since there is considerable communication activity in handling a fleet of taxicabs, frequency allocations are often made in pairs, one frequency for the basic station and a second frequency for the mobile units. Thus, the base station is able to hear all of the mobile units, but each mobile unit can only receive the base station. A mobile unit does not hear the other mobile units, and thus considerable two-way radio confusion is avoided. Radiocommunication can then be handled more efficiently and effectively between the base station and the mobile units.

Those eligible for licensing in the automobile emergency radio service are associations of owners of private automobiles who provide a private emergency road service for disabled vehicles, and those regularly engaged in the business of providing emergency road service for disabled vehicles to the general public. Communications must be limited to the safety of life or the protection of important property, and communications required for dispatching repair trucks, tow trucks, or other road-service vehicles to disabled vehicles. Also, associations of owners of private automobiles which provide emergency road service may transmit communications for the purpose of reporting traffic conditions on occasions of abnormal vehicular congestion.

Motor Carrier Radio

An active category of the Land Transportation Service has to do with motor carriers. Frequencies are available for persons engaged in providing a common or contract motor-carrier passenger-transportation service between urban areas, or for those who operate such a service within such an area. Frequencies are available for those persons primarily engaged in providing a common or contract motor-carrier property-transportation service between urban areas or for local distribution or collection of property. Trucking concerns are a principal user of these radio assignments.

Railroad Radio

Railroads have a variety of applications for two-way radio when regularly engaged in the transportation of passengers or property. Facilities may be used in connection with operation or maintenance including use in connection with motor vehicles engaged in the pickup, delivery, or transfer of property between stations. Since railroads stretch out over considerable distances, there is a need for relay as well as repeater stations.

In lieu of call letters for identification, the railroad radio services may use the name of the railroad and train number, caboose number, engine number, or name of fixed wayside stations. Unit identifiers may also be employed if they are kept on record.

CITIZENS BAND (CB) RADIO SERVICE

The Citizens Band (CB) Radio Service has had a phenomenal growth, as attested by the millions of transmitters now in operation. A segment of the frequency spectrum has been set aside for the Citizens Band Radio Service to provide a private, short-distance radiocommunications service for the business or personal activities of the licensees. Any person who is 18 years old or older may obtain a station license in this service if his application meets the requirements of the Citizens Band Radio Service. Partnerships, associations, trusts, or corporations meeting these requirements can be licensed including any state, territorial, or local government entity, or any service organization or association, includng civil defense. No operator license is required.

The Citizens Band Radio Service has many uses. It is particularly attractive because low-cost, efficient equipment is available, and almost everyone is eligible for a license. The Citizens Band Radio Service is basically mobile in nature with a majority of installations made in trucks and autos. Travel conditions, highway instructions, and personal exchanges dominate the mobile conversations. There are local activities as well as continuous chatter among trucks and private cars on the interstate highways.

The family base-mobile installation is particularly popular. The base station (still considered a mobile installation in terms of FCC terminology) is installed at home, and a mobile unit is installed in the family car. When one member of the family is shopping or traveling to or from work, that member is able to maintain contact with home. In this manner delays, travel changes, and additional shopping information can be confirmed and arranged via two-way radio. Often the time of departure from one's business or place of employment is quite indefinite. At a reasonable distance from home, a mobile-to-home contact can be made to let the family know said person is on the way home.

Of course, the two-way radio is handy in case of a car breakdown or other delays. In many localities it is not only possible to call home, but necessary assistance can be obtained by calling a station operated by a garage or service station. Channel 9 has been set aside for the exclusive use in requesting road help or for other emergencies. Citizens band radio is often quite helpful on long trips, because breakdowns on turnpikes often involve considerable delay if you have to await the routine, scheduled trip of a repair truck along the highway. Citizens band operation permits you to obtain a variety of services ahead of time, such as lodging, repairs, location and traffic guidance, medical assistance, etc.

Citizens band radio can be a special boon to a small businessman. Any type of pickup and/or delivery service can derive benefit in terms of reporting delays, breakdowns, changes in routing, and last-minute pickups or deliveries. A fuel-oil delivery service represents only one example of the many businesses that can use CB radio to advantage. Enroute trucks can be informed of call-ins from customers right after they have been completed and while the truck may be in the very vicinity of the customer's location. Parts jobbers in the service field use CB to advantage in making their deliveries to retail stores and service shops. Communications can even be maintained between the main store and outlying branches of various types of businesses.

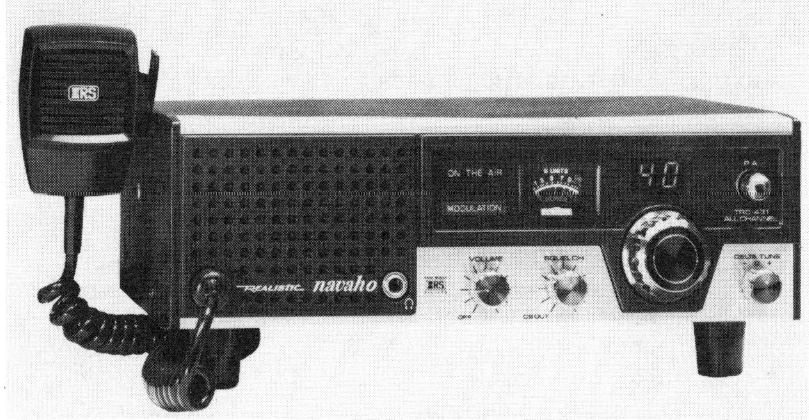

Courtesy Radio Shack

Fig. 2-24. CB base station.

Professional people can use citizens band radio for maintaining contact between car and office. Late call-ins or emergencies can be radioed to the doctor as he makes his rounds. Similarly, a veterinarian making his farm calls can keep in touch with his office.

There is quite a variety of citizens band radio equipment available. Base stations, shown in Fig. 2-24, are generally designed to operate from 120 volts ac while the mobile units (Fig. 2-25) operate from the 12-volt dc car battery. Some transceivers can be used with either type of primary input power. Still other equipment is of the hand-held type (Fig.

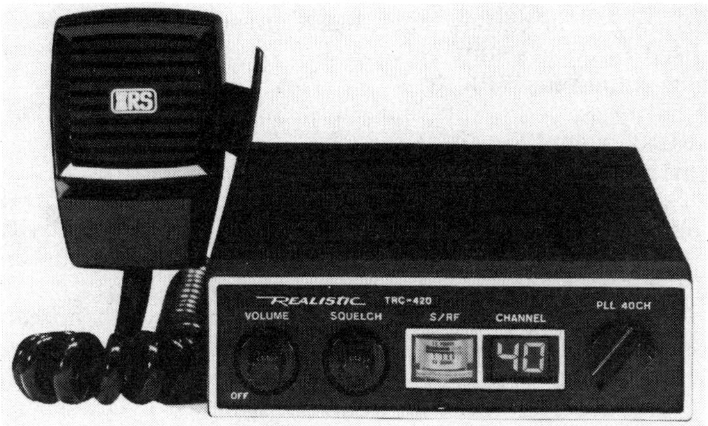

Courtesy Radio Shack

Fig. 2-25. Mobile CB transceiver.

Fig. 2-26. Hand-held CB transceiver.

Courtesy E. F. Johnson American Inc.

2-26), which is battery operated. Conventional batteries can be used in such a unit or the more recent rechargeable type can be employed. With the latter type, it is possible to recharge the batteries overnight using a small charging unit that is operated from 120-volt ac power.

Citizens band radio can have many public service functions, such as being used for liaison work for trade shows, fairs, or other large private or public events. Emergency activities, traffic control, parades, sporting events, etc., can often be expedited with two-way radio gear.

In an office or a small business situation a CB transceiver installation can be made less annoying and with more privacy if it is supplied with a radiotelephone handset, such as shown in Fig. 2-27. This same unit can be switched over to normal loudspeaker operation when desired.

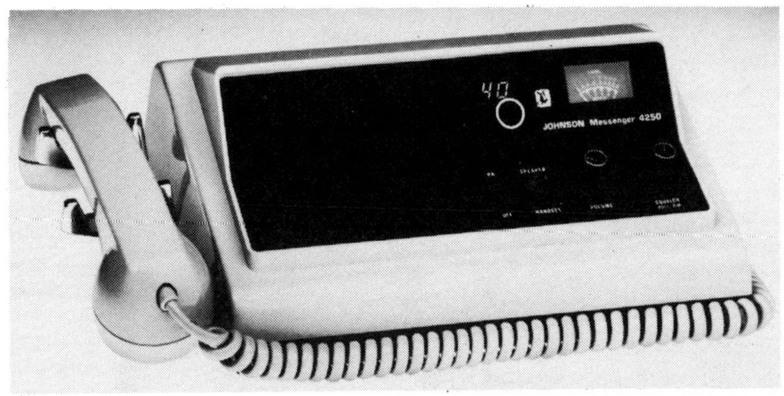

Courtesy E. F. Johnson American Inc.

Fig. 2-27. A CB base station with handset.

Single-sideband emission has become increasingly popular on the citizens band. Usually, the transceivers (Fig. 2-28) include standard amplitude modulation plus single-sideband capability. When using standard amplitude modulation, the maximum permissible carrier power is 4 watts and the emission from the transmitter occupies the band of frequencies shown in Fig. 2-29A. Note that there is a span of frequencies on each side of the carrier frequency that extend above and below the carrier by an amount equal to the highest audio frequency. A practical value for this upper audio frequency for voice communications is 2500 hertz. Thus, as shown in Fig. 2-29A, the emission occupies a span of frequencies equivalent to 5000 hertz.

Courtesy Radio Shack

Fig. 2-28. An am/single-sideband CB transceiver.

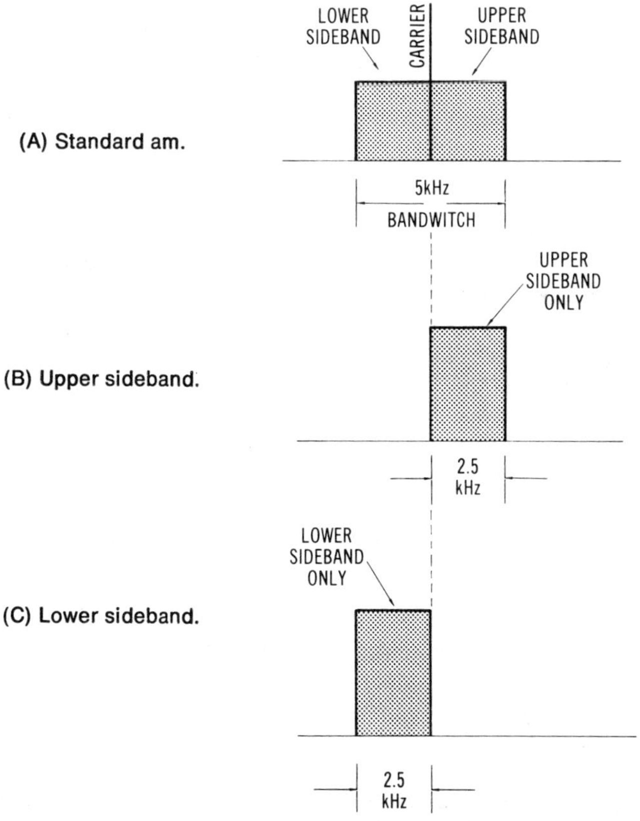

(A) Standard am.

(B) Upper sideband.

(C) Lower sideband.

Fig. 2-29. Standard am and single-sideband characteristics.

When sideband emission is used, the power permitted is 12 watts peak of single-sideband signal. Furthermore, depending upon whether upper or lower sideband emission is used, the span of frequencies occupied by the signal is only about 2500 hertz, as shown in Figs. 2-29B and C. Thus, more stations can be accommodated in the 40-channel spectrum allocated for CB operation. Furthermore, the effectiveness of the signal is improved and the range of operation is greater than that of standard am transmission. Of course, there is an additional cost involved when you purchase a unit that is capable of both standard am and ssb transmission.

Such a transceiver gives you much greater versatility. For example, on any given channel you can transmit a standard am signal or you have a choice of transmitting either a lower-sideband or an upper-sideband ssb signal. There can be interference between am and ssb stations and, therefore, by gentleman's agreement certain of the 40 channels are used almost exclusively for single-sideband transmission.

Citizens band operation need not be confined to land transportation. Transceivers can be installed in aircraft or boats. Furthermore, contacts can be made with base stations as well as other types of mobile stations. For example, an aircraft unit can communicate with a unit installed in a car moving along the highway, or a contact can be established between a small pleasure boat and a car driving along the shore.

A CB radio requires a station license. No operator license is needed although the licensee must know and abide by the FCC rules and regulations for CB radio operation. He must have a copy of the current rules and regulations. Complete rules and regulations are given in Appendix D. Study them carefully to obtain the required knowledge about CB operations and procedures.

GENERAL MOBILE RADIO SERVICE

Like the Citizens Band Radio Service, the General Mobile Radio Service allocations of the Personal Radio Services provide for private short-distance personal or business radio communications. These stations operate on assigned frequencies in the 460–470-MHz band with a transmitter output power of not more than 50 watts. Frequency assignments are given in Table 2-3.

Those frequencies in the left column can be assigned to either base or mobile stations. The frequencies in the right column are assigned to mobile stations only. These are known

Table 2-3. Frequency Assignments for the General Mobile Radio Service

Base and Mobile	Mobile Only
462.550	467.550
462.575	467.575
462.600	467.600
462.625	467.625
462.650	467.650
462.675	467.675
462.700	467.700
462.725	467.725

as *paired-frequency* assignments with the mobile stations transmitting on frequencies that are 5 megahertz higher than the base transmitter frequency. In effect there are eight General Mobile channels, each consisting of two frequencies. In a simple duplex system, Fig. 2-30, the mobiles transmit on a frequency 5 MHz higher than the base station. For example, if you receive a base-station transmitter frequency assignment of 462.7 MHz you will transmit on this frequency and listen for your mobile stations on 467.7 MHz. Conversely the mobile units will transmit on 467.7 MHz and receive on 462.7 MHz. This is known as a paired-frequency channel.

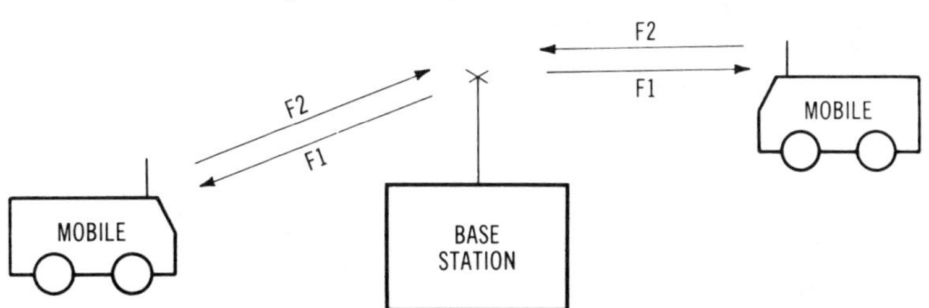

Fig. 2-30. Paired-frequency operation in the General Mobile Radio Service band.

However, it is also possible to set up a system with only a single frequency which must be selected from those in the left-hand column. In this case both base and mobile stations operate on the same frequency; this is known as *simplex operation*.

Typical base and mobile installations are shown in Fig. 2-31. These crystal-controlled units operate on assigned frequencies between 462–468 MHz. They are solid-state transceivers with

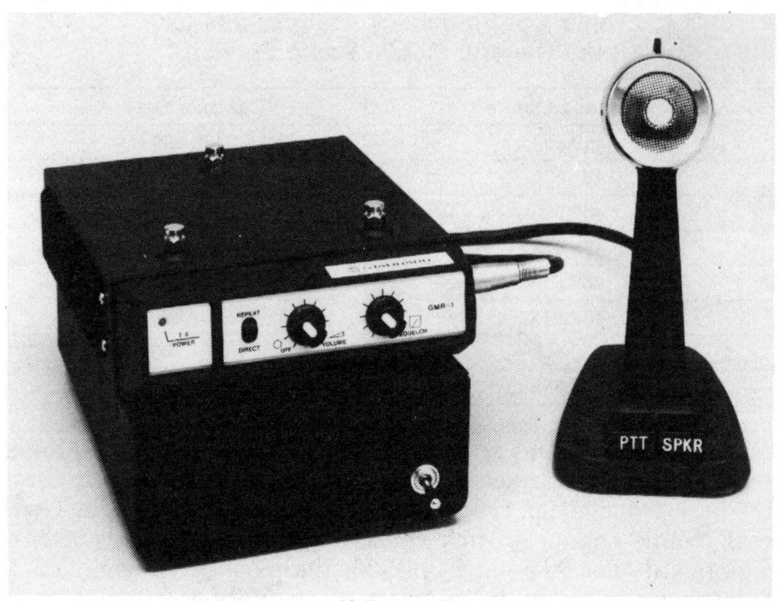

(A) Base unit.

(B) Mobile installation.

**Fig. 2-31. Base and mobile installations for the General Radio Mobile
Service band.**

a power output of approximately 6 watts. Two-channel operation is also possible.

For a typical installation, the FCC assigns only one of the frequencies in the left column to the licensee. This becomes the base-station transmit frequency. Mobiles can operate on this same frequency or the assigned mobile frequency for that particular paired-frequency channel. In this situation communications can only be established on these frequencies unless additional receive frequencies can be accommodated by the transceiver. Although a base station may only transmit on one assigned frequency, there is no restriction on being able to receive other frequencies provided that the appropriate receive crystals are installed. Thus, it is possible to have cross-channel operation. For example, if your transmit frequency is 462.7 MHz but your receive frequency is 462.725 MHz, which is the same as the transmit frequency of another base station, it is possible to establish contact between the two base stations provided both installations have the correct receive frequency for cross-channel operation.

The General Mobile Radio Service is in the uhf band where ignition noise is at a very minimum. Furthermore, there is no skip interference. The General Mobile transceivers must use frequency modulation resulting in a further reduction in noise level. Also, there is a minimum of interference among stations operating on the same frequency because the stronger signal always dominates at a given receive position. The maximum power permissible is 50 watts, the effectiveness of which can be increased with a high-gain antenna. Remember, at this frequency, a high-gain antenna is very practical because of its small size.

The range of transmission falls between 10 and 20 miles, depending upon terrain as well as the gain and height of the base-station antenna. The transmission range can be extended with the use of a repeater station which is permissible in the General Mobile Radio Service. This repeater can be located at the top of a mountain or a tall building and, if high enough, could possibly extend the range of transmission by a factor of 3 or 4. In this type of installation the base station would act as the control station with its antenna directional and pointed toward the repeater. All communications with the mobile units would be by way of the repeater.

It is also permissible to have shared operation of the repeater. In this arrangement several licensees would share the cost of operating the repeater station. All would benefit because of the high location of the repeater station antenna.

CHAPTER 3

Radio Broadcast Operation

In recent years holders of licenses lower than first-class radiotelephone (broadcast technician) are being employed as duty operators in am and fm radio broadcast operations. The development of more stable and troublefree transmitters has made this possible. At some stations completely automatic transmitter systems are in use.

If you have a restricted operator permit, you can be instructed in the duties of transmitter monitoring under the supervision of the first-class license holder. This restricted license is available by application and requires no examination. One of the attractive facets of the FCC licensing program is that you can become a technical person in radio broadcasting without involvement in the time and cost of obtaining a baccalaureate degree. You can gain experience and knowledge as you study for the advanced license examinations.

The restricted permit holder can also serve as a duty operator in standard am stations that use directional antennas. The broadcast technician license is mandatory only when the am station uses an antenna with a critical directional pattern that is not automatically controlled. The restricted permit holder can also serve as a duty operator in fm broadcast stations, commercial and educational. It is expected that such a permit will be adequate for almost any type of automatic transmission system.

It is wise, however, to at least have some idea of the basic makeup of a broadcast station and the duties and responsibilities of the license holder even though the license required might be only a restricted permit. The objectives of this chapter are to give you some background in radio broadcasting, both technical and, especially, with regard to compliance with the FCC Rules and Regulations.

TECHNICAL CONSIDERATIONS

The Federal Communications Commission is concerned with the performance of the transmitter and the technical characteristics of the signal that is radiated from the broadcast antenna. Certain strict technical requirements have been set down with regard to this radio signal so that it may be used effectively by the receivers tuned to its frequency. This broadcast signal must not interfere unduly, within the state of the science, with stations operating on other frequencies. Likewise, its radiation must be such that it does not interfere, within certain established interference ratios, with stations in other areas operating on the same frequency. Hence, the radiated power output of the broadcast station must be held within the power limits specified by the FCC. If the station uses a directional antenna, the radiation pattern produced by that antenna must conform with specified FCC limits.

Each broadcast station must operate on its assigned frequency with a very tight tolerance. Broadcast channels are closely spaced and the broadcast-carrier center frequency must not drift so as to interfere with stations operating on adjacent channels.

The voice or music signal that is applied to the radio-frequency wave must be added in an efficient and correct manner. Stated technically, the radio-frequency carrier must be properly modulated by a voice or music signal. When a radio-frequency carrier is modulated fully, it is said to be 100-percent modulated. Only the strong, peak audio signals modulate the carrier to this extent. The average content of voice or music signal may modulate the radio-frequency carrier by approximately 70 percent.

If the voice or music components are too strong, in comparison to the strength of the radio-frequency carrier, the transmitter is said to be *overmodulated*. When the transmitter is overmodulated, the signal, as it is recovered by a radio receiver tuned to that station frequency, may be distorted. An overmodulated transmitter also generates what are called *spurious signals*. These spurious signals appear on frequencies other than the one which has been assigned to the station.

Thus, overmodulation of the radio-frequency carrier can cause interference in the reception of broadcast stations using other frequencies.

A transmitter can also be *undermodulated*. In this case the radio carrier is too strong in comparison to the audio content that it is to convey. Under these conditions, the broadcast station does not attain its maximum range of transmission and the signals sound weak in all but those locations that are relatively close to the broadcast antenna.

In summary, the FCC imposes strict requirements in terms of the strength of a signal radiated from the antenna, the frequency of the radiated signal, and whether or not this radio carrier is properly modulated.

THE OPERATOR'S RESPONSIBILITIES

The meters and indicators with which the broadcast operator is concerned provide a visual indication of just how well the broadcast transmitter is meeting the specified technical requirements. Thus, the surveillance of those meters is of significance both in terms of the quality of the broadcast signal and the compliance of the station with the FCC technical standards.

The broadcast duty operator must keep a routine watch on operating conditions. In fact, a regular log of operating readings must be kept, manually or automatically, to ensure compliance with the FCC rules and regulations. If the readings do not meet or stay within certain limits, appropriate corrections must be made, or in extreme cases, the transmitter must be shut down and the necessary repairs and adjustments made.

Proper attitudes toward technical operating responsibilities are very important, since improper operation may bring an FCC citation or, in the extreme case, result in a fine or loss of license.

The normal operating position for the restricted operator may be at the transmitter, at the remote control point, or at the extension meter location. Such an operator is authorized to make the following adjustments: (1) turn the transmitter on and off, (2) compensate for voltage fluctuations in the primary power supply, (3) maintain modulation levels within prescribed limits, (4) effect routine changes in operating power, and (5) change antenna patterns. The operator must post either his permit or Form 759 at the transmitter, the remote control point, or the extension meter location.

To prevent or minimize skywave interference, many am stations are required to cease operating, reduce power, or change their antenna pattern at sunset. These stations return to their daytime operating conditions at sunrise. The station licensee must post printed instructions and a chart of operating parameters for the guidance of restricted operators at these stations.

TRANSMITTER METERING

There are several key transmitter monitoring meters; two of these meters are the *final-plate-voltage meter* and the *final-plate-current meter*. The final plate voltage must be of correct value for operation of the stage that generates the final high-power radio-frequency carrier. The normal voltage is usually several thousands of volts. If the voltage is too low, the final stage will not generate a strong enough rf carrier. If the final plate voltage is too high, power output may be too great or the associated equipment may be damaged.

The final-plate-current meter indicates how much current the final rf power amplifier is drawing. It is an indication of how well this stage is operating, and whether or not it is supplying the proper level of power to its output circuit. Most transmitters have provisions for adjusting these quantities so that operation of the transmitter may be set at some point to provide optimum plate voltage and plate current.

A third meter is the *power-output meter*. It measures the power that is being transferred from the final stage of the transmitter to the radiating antenna, via the transmission line. This output is read on an rf antenna-current meter, or a calibrated indicator called a reflectometer.

The output-meter reading is very important because it tells just how well the rf power being generated by the transmitter is transferred to the antenna, and how well the antenna system is operating. Depending on system design, the actual reading may change with weather and moisture conditions. Adjustments can usually be made to compensate for weather and terrain effects.

Another important meter is the *modulation meter*. This meter indicates how effectively the voice or music components are modulating the rf carrier. The calibration is given in percentage, indicating 100 percent when the carrier is being fully modulated. A lower level of modulation is indicated by a lower percentage reading. If the voice or music components are made too strong for proper modulation of the rf carrier, the meter will indicate a modulation percentage in excess of 100 percent. Often a flasher or clacking relay will indicate when the transmitter is being overmodulated.

Corrective action must be taken by the operator when there is excessive modulation. Generally, the modulation should not be less than 85 percent on peaks of frequent recurrence. However, it may be less than 85 percent when necessary to avoid objectionable loudness. At am broadcast stations, modulation must not exceed 100 percent on negative peaks and 125 percent on any positive peaks. At fm broadcast stations, modulation must not exceed 100 percent on either positive or negative peaks.

The antenna ammeter is inserted to measure current into the antenna system. This meter is a direct indication as to whether the station is operating at, above, or below the licensed power. The antenna current should be maintained as close as possible to the licensed operating current, and the operator should know what actions are necessary when the meter indications deviate from that value.

Usually, the antenna ammeter is located at the base of the tower and is not easily accessible to the operator on duty. For this reason, most stations use a remote antenna ammeter that is located at the normal operating position of the person on duty. The remote antenna ammeter indication may be entered in the operating log in lieu of the ammeter indication at the base of the antenna provided that the remote meter is calibrated weekly.

Nondirectional am stations use a single antenna tower and transmit the radio signal with equal strength in all directions from the station. Directional am stations utilize more than one antenna tower. By establishing the position of each tower, the power radiated by each tower. and the phase of the signal in each tower, different signal strengths can be radiated in various directions. Directional antenna systems are used to improve the signal over desired areas, and to reduce the signal in the direction of other stations to prevent interference. The operator on duty at a directional am station must monitor the tower currents as well as the phase angles among the tower currents. To determine if a directional antenna system is radiating the signal according to a specified radiation pattern, an instrument called an *antenna monitor* is installed at the station. The antenna monitor enables the operator to determine if the radio-frequency current in each tower is of the correct value and if the phase of the signal radiated by each tower is also correct. Some antenna monitors indicate the ratio of current in each tower to the current in one tower called the *reference tower*.

If the signal arrives at each tower at the same time, the current in each tower is said to be in phase. In most directional antenna systems, the time that the radio-frequency signal reaches each tower from the transmitter is not the same. The rf wave phasing or time difference is measured in degrees. The station license contains a list of the required signal phases and antenna base-current ratios for all the towers in the directional antenna system.

The antenna base-current ratio for a tower is calculated by dividing the current-meter reading for that tower by the current-meter reading for the designated reference tower. The ratio, either calculated or read on the antenna monitor, must not deviate more than 5 percent from the value on the station license. (For some antennas, the ratio is extremely critical, and a closer tolerance is stated in the license.)

Operators on duty at am stations using directional antenna systems should know how to read the antenna-monitor meters. They should also know how to use charts, tables, or other instructions to determine if the station is operating correctly, if attention by the first-class operator of the station is required, or if the station must terminate operation.

Typical meter faces and readings are shown in Fig. 3-1. Plate-current and plate-voltage meters are shown at the top, indicating a plate-current reading of 270 milliamperes and a plate voltage of 3200 volts.

The antenna-current meter is reading 1.7 amperes and the power-output meter is reading 4.5 kilowatts. The modulation meter is indicating 75 percent modulation. The antenna phase meter shows a reading slightly higher than 24 degrees.

OPERATING POWER

Each am and fm broadcast station is authorized to operate at a specified operating power as indicated on the station license. The operating power must be maintained as near as possible to the value specified by the station license and shall not be more than 105 percent nor less than 90 percent of this level. Noncommercial educational fm broadcast stations licensed to operate with transmitter output power of 10 watts or less may be operated at less than the authorized power, but not at more than 105 percent of the authorized power.

Nondirectional am broadcast stations employ a single antenna tower. Power determined by the direct method is equal

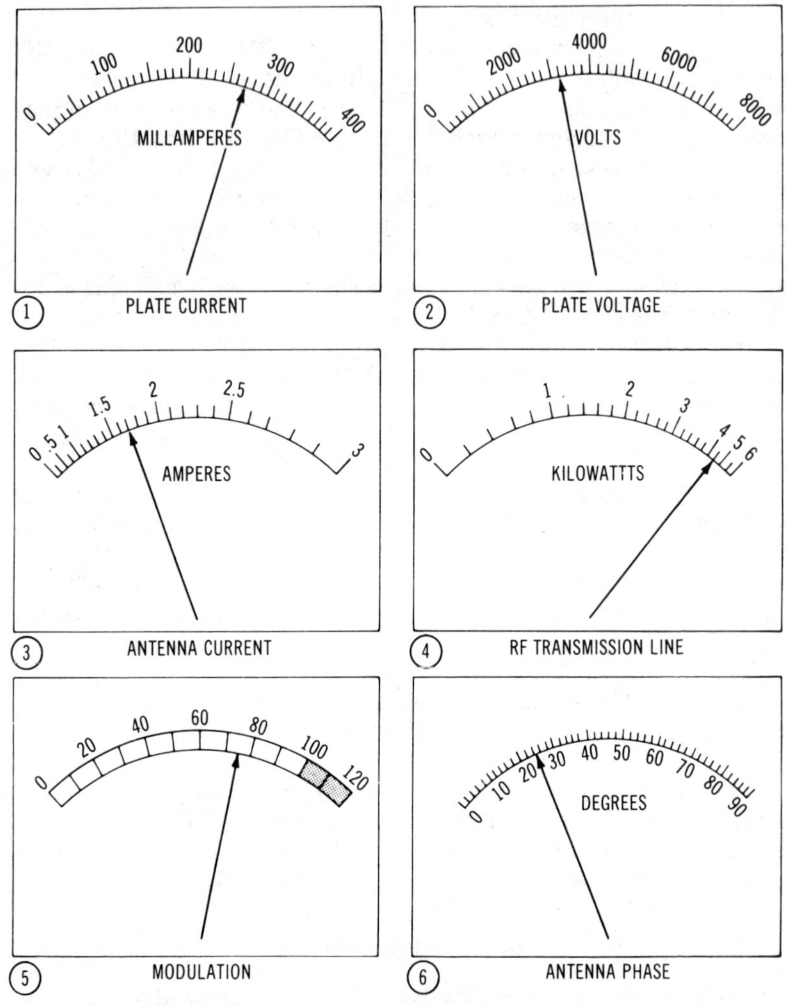

Fig. 3-1. Typical meter faces.

to the product of the antenna resistance and the square of the antenna current:

$$\text{Operating Power} = I_A^2 R_A$$

Directional am stations employ multiple radiating elements. Power determined by the direct method for these stations is equal to the product of the resistance common to all antenna towers (called the common-point resistance) and the square of the current common to all antenna towers (called the common-point current).

At fm broadcast stations, if the power is determined by the direct method, it is read directly from the rf transmission-line meter. This meter must be calibrated each six months to give a direct indication of the power supplied to the antenna.

Generally, am broadcast stations must determine the operating power by the direct method. The operating power of fm broadcast stations may be determined by either method, but the indirect method is used more.

For both am and fm broadcast stations, operating power determined by the indirect method is equal to the product of the plate voltage and the plate current of the last radio stage, and an efficiency factor:

$$\text{Operating Power} = E_P I_P F$$

When the power of an am station is determined by the indirect method, the calculated value is to be entered in the operating log when required transmitter readings are taken.

In summary, the operator on duty is responsible for monitoring the station operating power which is measured by either the direct or indirect method. In the direct method, measurements are made at the base of the antenna (am), the common point (am directional), or in the transmission line (fm). With the indirect method, measurements are made at the input of the final amplifier.

At all stations where restricted operators are employed, the upper and lower limiting values must be posted for antenna current, common-point current, transmission-line meter values, plate voltage, and plate current.

At some stations key readings are monitored and/or recorded automatically. If there are improper operating conditions, automatic warnings are given. The duty operator must respond to these warnings in accordance with the instructions given by the first-class operator or chief engineer in charge.

STATION PLANS

If a station is to be made acceptable for operation by the holder of a lesser grade license, the various indicating meters and operating controls must be made accessible to the operator at his normal on-duty position. Most often his duties involve work as an announcer and/or control-board operator.

A typical station arrangement is shown in Fig. 3-2. In this plan the entire station is incorporated into a compact area.

The key transmitter readings can be observed when the front panel of the transmitter faces the control console. The equipment rack to the left of the operating position houses the monitoring equipment. From his operating position the con-

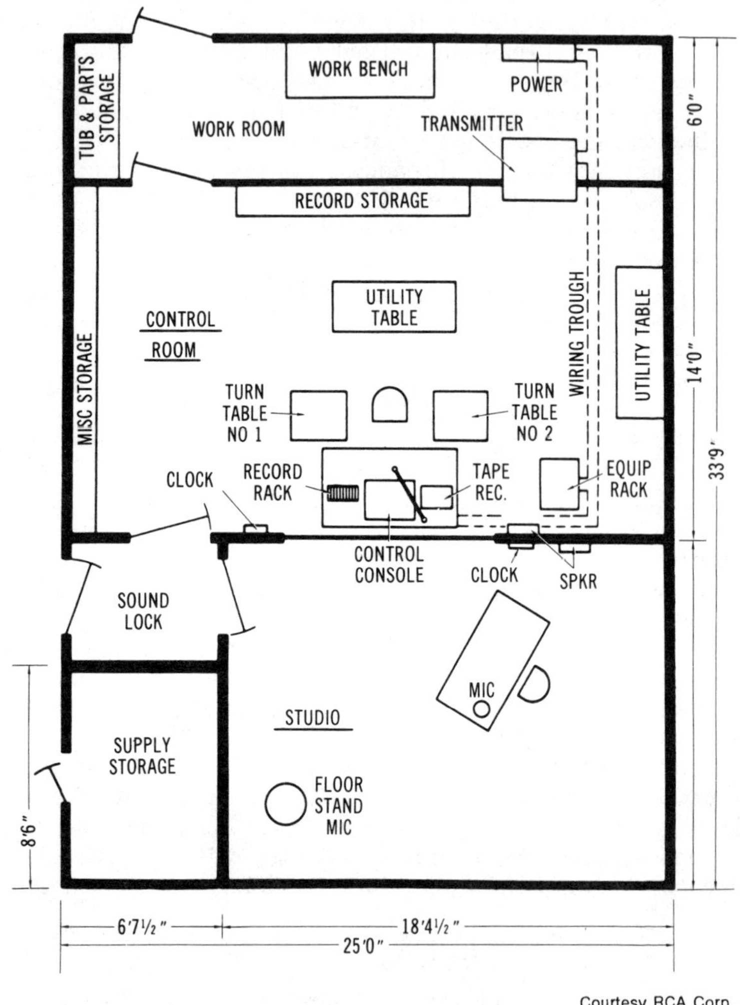

Courtesy RCA Corp.

Fig. 3-2. Basic station plan.

trol-room operator also can look through a window directly into the studio.

Actually, there are three basic acceptable arrangements. In some installations the transmitter is installed in such a manner that it is visible through an appropriate window. The duty operator can keep a watch on transmitter meters from the operating position at the control console. The modulation monitor can be located in an equipment rack near the operator.

A second acceptable arrangement has the control-console position looking into the studio. However, the transmitter proper is located in an adjacent room and is not visible to the operator on duty. A duplicate set of key meters is mounted in an equipment rack near the operator. These meters are connected via wires to the metering circuits of the transmitter. The equipment rack also usually houses the modulation monitor.

A third acceptable arrangement uses a remote-control facility. In this arrangement the transmitter may be located at a site that is quite distant from the control-console operating position. However, at the operating position there will be a remote-control panel which can be used to control transmitter circuits and operation over interconnecting telephone lines. In this case, appropriate metering facilities that can be used to evaluate the transmitter performance from the operating position are installed in the control room. Likewise, suitable switches and controls are connected to permit the required transmitter adjustments to be made remotely. FCC rules for remote control operation are as follows:

Operation by remote control shall be subject to the following conditions:

(1) The equipment at the operating and transmitting positions shall be so installed and protected that it is not accessible to or capable of operation by persons other than those duly authorized by the licensee.

(2) The control circuits from the operating positions to the transmitter shall provide positive on and off control and shall be such that open circuits, short circuits, grounds or other line faults will not actuate the transmitter and any fault causing loss of such control will automatically place the transmitter in an inoperative position.

(3) A malfunction of any part of the remote-control equipment and associated line circuits resulting in improper control or inaccurate meter readings shall be cause for the immediate cessation of operation by remote control.

(4) Control and monitoring equipment shall be installed so as to allow the licensed operator at the remote-control point to perform all the functions in a manner required by the Commission's rules.

Many broadcast station ownerships include both am (535–1605 kHz) and fm-stereo (88–108 MHz) outlets. One such successful operation in Philadelphia, Pennsylvania is WFIL (5 kilowatts on 560 kHz) and WUSL (14 kilowatts on 98.9 MHz). A WFIL studio control room looking into one of several studios is shown in Fig. 3-3. The switching group and fader controls that permit the intermixing and volume-level control of the various sound sources are shown at the center. Cartridge tape players or "cart players" are mounted on top of the operating deck and phonograph turntables are located at each side of the operating position. A reel-to-reel tape player is shown at the left.

Courtesy WFIL

Fig. 3-3. Modern control room for an am broadcast station.

A remote monitoring and control rack, shown in Fig. 3-4, is a part of the same room. Switching of the remote monitoring facility permits a digital readout display of key transmitter and antenna parameters. The remote control function at this point permits adjustment of key transmitter parame-

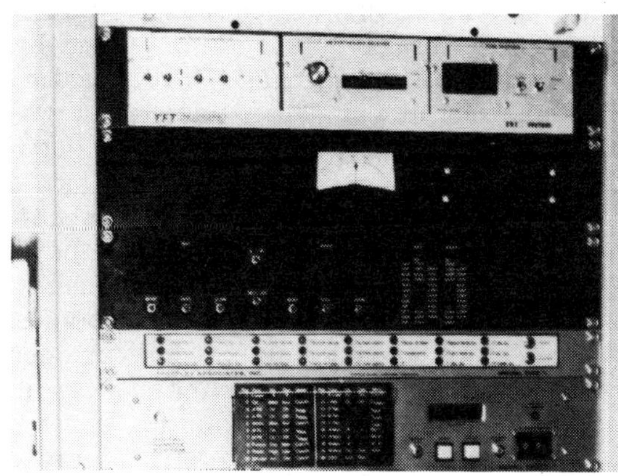

Fig. 3-4. Remote-control and monitoring rack for am station WFIL.

ters even though the actual transmitter site is miles away. WFIL has a directional transmitting pattern and the patterns themselves differ for daytime and nighttime operations. The

Fig. 3-5. Extension metering rack for fm station WUSL.

transmitter log is kept at this remote-control position in the studios.

The rack also mounts the Emergency Broadcast System receiver which must be in continuous operation. WFIL and WUSL also transmit EBS test signals on a random schedule.

The studios of the fm station, WUSL, are located at the transmitter site in another part of the city. Transmitter monitoring is the responsibility of a so-called "combo" operator. The operator has charge of the program sequencing and announcing as well as being responsible for monitoring key transmitter parameters. To monitor transmitter operation, he is supplied with an extension metering and monitoring facility (Fig. 3-5). Information about modulation, transmitter output, and final transmitter voltage and current are presented by the monitoring facility. In case of a transmitter failure, it can be shut down from this position by the "combo" operator. The "combo" operator must have a minimal license grade of restricted radiotelephone operator permit.

WGSA-WIOV

This am/fm combination plant is progressive and modern. Its primary service area is the small community of Ephrata,

Fig. 3-6. Operating position for "line" programming at WIOV. Automatic programming panels are shown in the background.

Pennsylvania and environs. However, both stations are received well in the nearby small cities of Reading and Lancaster and along a good stretch of the Pennsylvania Turnpike. In fact, the fm stereo outlet, WIOV, "live" broadcasts a truckers' and travelers' program from midnight until 9 am. The operating position for this "live" program is shown in Fig. 3-6. Program automation takes over for the remainder of the 24-hour day. The country-western program format is set up the evening before. Automated cartridge-racks, computer, and control point are shown in Fig. 3-7. Commercials

Fig. 3-7. Cartridge racks, computer, and control point for program automation.

and other material are taped on cartridges and inserted into the program sequence. The daytime am station, WGSA, is also automated for most of the day.

The fm transmitter and antenna system with an effective radiated power of 50 kilowatts is located on a nearby ridge. It is operated by remote control. Remote-control unit, stereo monitoring, and SCA monitor are shown in Fig. 3-8. This control and monitoring rack can be seen by the am transmitter operator and from the operating position shown in Fig. 3-6.

The fm transmitter logging sheet is shown in Fig. 3-9. Note that under the operating constants are listed time, frequency deviation, final plate current, final plate voltage, and percentage power output. There are separate carrier and program on-and-off times that must be logged for stereophonic and SCA operation.

WIOV also broadcasts sporting events using a uhf air-link to beam the program to the fm transmitter site. From here the signal is sent over telephone lines down to the main operating position where the audio material is coordinated and placed on the program line back to the transmitter.

Logging is the responsibility of the transmitter operator. The operator of the automation equipment must be a skilled person. Although this person does no announcing during automatic programming, he must keep continuous watch on operating sequence and set-up. He also must have a knowledge of computer operation.

The am transmitter is a 5000-watt unit and supplies signal to a three-tower directional antenna system. The directional antenna monitor that indicates phase angle and loop currents is shown in Fig. 3-10. Although the am station operates much of the day on program automation, there is also a studio and a separate control room for this outlet.

Fig. 3-8. Remote-control. stereo monitoring, and SCA monitor units for fm transmitter.

The WGSA am transmitter log is shown in Fig. 3-11. Note the various readings that must be logged, in particular, the six antenna measurements that must be recorded in the log. The transmitter operator keeps the logs for both the am and fm outlets.

All broadcast stations must incorporate an Emergency Broadcast System receiver (Fig. 3-12). If the station also transmits an emergency signal according to the scheduling of the Emergency Broadcast System, it must also include the appropriate tone generator.

MONITORING SUMMARY

Key monitoring and metering facilities of a broadcast transmitter are shown in Fig. 3-13. As a part of the transmitter proper, there are two meters that read the final current and final voltage. The product of these meter readings is the input power of the transmitter. These two meters are compulsory.

Associated with the transmitter is a remote common-point antenna-current meter. This meter provides a remote indication of the common-point antenna current being supplied to the phasor which is a part of the directional antenna system. This current squared times the resistance of the antenna system at the phasor input is a measure of the transmitter power output. The common-point antenna-current meter is also compulsory.

A compulsory modulation meter associated with the transmitter determines the degree to which the audio modulating wave modulates the transmitter output. A frequency meter is used to determine that the transmitter is operating on the FCC-assigned frequency. In modern stations, this meter is common but often is not logged because of the inherent frequency stability of modern transmitters.

The function of the phasor is to properly divide the currents supplied to the two antennas of a two-tower directional antenna system. The relative magnitude of the two currents as well as the timing (phase relationship) between the two currents is important in establishing a proper antenna pattern. Meters are used to monitor the individual antenna leg currents.

A directional antenna system is employed at many stations to make certain that the transmitted signal does not interfere in the reception areas of broadcast stations operating on the same frequency in some distant city. A directional pattern also helps to concentrate the energy into the areas that are to be

RADIO STATION

50,000 WATTS E.R.P.
HORIZONTAL & VERTICAL

105.1 MHZ

EPHRATA, PA.

DATE:

DAY:

WIOV

OPERATING CONSTANTS					STEREOPHONIC OPERATION	
Time EST EDT	Frequency Deviation Hz	Pa Plate I Amps	Pa Plate E Volts	% Power Output	Sub Carrier On	PGM on
					Sub Carrier Off	PGM off
					Sub Carrier On	PGM on
					Sub Carrier Off	PGM off
					Sub Carrier On	PGM on
					Sub Carrier Off	PGM off
					Pilot Frequency Deviation: _______ Hz at_____ _______ Hz at_____ _______ Hz at_____ _______ Hz at_____	
					SCA OPERATION (67 KHz)	
					Sub Carrier On	PGM on
					Sub Carrier Off	PGM off
					Sub Carrier On	PGM on

Fig. 3-9. Log sheet for

fm transmitter .

Sub Carrier Off	PGM off
Sub Carrier On	PGM on
Sub Carrier Off	PGM off

Sub Carrier Freq. Deviation: __________ Hz at ____
__________ Hz at ____
__________ Hz at ____
__________ Hz at ____

Carrier On: __________
Program On: __________
Program Off: __________
Carrier Off: __________

Carrier On: __________
Program On: __________
Program Off: __________
Carrier Off: __________

OPERATOR'S SIGNATURE

On __________ Off __________   Tower Light Checked: __________
On __________ Off __________
On __________ Off __________   Log Checked: __________
On __________ Off __________

covered by the radio station. A compulsory antenna monitor is needed to evaluate the phase between the two currents as well as the relative magnitudes. When the operating position is not at the transmitter, an extension monitoring facility must be set up at the operating position.

If the transmitter is remotely controlled, a duplicate monitoring arrangement must be employed at the operating position. In a remote-control facility there are also the necessary controls needed to make the proper transmitter adjustments to insure compliance with the FCC Rules and Regulations.

Fig. 3-10. Directional antenna monitor.

REMOTE CONTROL

A remote-control metering panel is shown in Fig. 3-14. This panel is located in the studio or control-room of the station. A second unit of the system is installed at the transmitter and makes the necessary interconnections that permit the transmitter to be controlled remotely in accordance with the indications of the remote metering facility. The control point and transmitter are interconnected using telephone lines or an actual studio-to-transmitter radio link. The three meters permit the measurement of final voltage, final current, and antenna current. The transmitter remote-control facility gathers the information from the appropriate transmitter circuits and places a calibrated voltage on the link.

In the arrangement shown in Fig. 3-14, the three meters do not record simultaneously but are switched into the circuit by a multiple-position switch. Another switch position gives indication of tower-light circuit operation. Various other switch positions are used in a variety of control, monitoring, and metering applications.

A directional antenna monitor keeps a check on antenna phasing and current levels among the towers of a directional antenna system. The directional antenna monitor in Fig. 3-15 employs a digital readout. Up to six towers can be monitored by depressing the appropriate button. The antenna current relative to the zero-angle reference is shown by the left readout. The right readout is the current ratio of a given tower relative to the current supplied to the reference tower.

Many modern stations employ digital remote control operation. Still others are progressing to completely automatic transmission systems. A conventional remote-control system monitors transmitter and station operation and then gives an indication when there is a malfunction. Manual activation of the remote-control adjustments is then made by an operator. In a completely automatic transmission system the corrections are made automatically and no operator need initiate the correction sequence.

A digital remote-control system comprises the control system at the studio or operating point (Fig. 3-16) and a second unit installed at the transmitter which is used for the actual metering and control of transmitter parameters. The link can be by means of telephone lines or a studio-to-transmitter microwave link.

The TFT system shown in Fig. 3-16 is basically a 10-channel digital system with raise/lower functions and 10 channels of telemetry information. Individual channels are selected using a thumb-wheel switch. Once selected, the channel number is fed back to the control point and displayed on the front panel for verification. The appropriate meter reading is then displayed in digital form by the readout. System meter readings are updated three times per second. Control commands can be initiated by pushing the appropriate raise/lower button on the front panel.

An automatic transmission system (ATS) permits transmitter operation under a completely automatic control system that monitors, controls and corrects, sounds alarm, and shuts down the transmitter when necessary. Key operating condi-

RADIO STATION

5000 WATTS D D

1310 KHZ

W G S A

EPHRATA, PA.

DATE:

DAY:

OPERATING CONSTANTS

TIME EST EDT	Frequency Deviation Hertz	Final Ip. Amps.	Final Ep. Volts	Common Point Current	Phase ° Ant 1 : 2	Phase ° Ant 3 : 2	Sampling Loop Ant. 1	Sampling Loop Ant. 2	Sampling Loop Ant. 3

Fig. 3-11. Log sheet for

ANTENNA SYSTEM OPERATION

Base Currents

Ant 1 _______ Amps

Ant 2 _______ Amps   Ratio 1:2 _______   Deviation 1:2 _____ %

Ant 3 _______ Amps   Ratio 3:2 _______   Deviation 3:2 _____ %

Remote Currents

Ant 1 _______ Amps

Ant 2 _______ Amps   Ratio 1:2 _______   Deviation 1:2 _____ %

Ant 3 _______ Amps   Ratio 3:2 _______   Deviation 3:2 _____ %

Common Point Current _______________ Calibration Time _______________ Signature _______________

OPERATOR'S SIGNATURE

On _______________   Off _______________   Carrier On: _______________

_______________   Off _______________   Program On: _______________

On _______________   Off _______________   Program Off: _______________

On _______________   Off _______________   Carrier Off: _______________

Log Checked: _______________

Fig. 3-12. Emergency Broadcast System receiver and tone generator.

tions related to modulation and power are monitored and, if necessary, automatic adjustment is made. If this procedure doesn't correct the problem or if certain ATS functions fail,

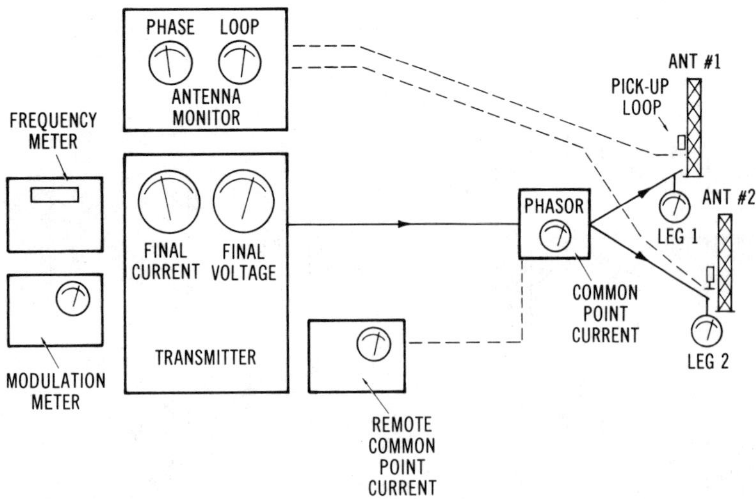

Fig. 3-13. Transmitter metering and monitoring.

Fig. 3-14. Remote metering panel.

the system sounds an alarm or shuts down the transmitter depending upon the nature of the failure.

Although an operator must be on duty, such a person need only have a restricted radiotelephone permit. However, with a restricted permit, the operator cannot perform maintenance functions.

STATION REQUIREMENTS AND OPERATOR DUTIES

When duty operators hold a lesser-grade radiotelephone license, certain station requirements must be met. Except at times when the station is under the immediate supervision of an operator holding a first-class radiotelephone license, adjustments of the transmitter equipment by the lesser-grade radiotelephone permit holder is limited. The holder of any grade of commercial license or permit may operate any class of standard am, fm, or educational fm broadcast station except those using directional antenna systems which are required by the station authorization to maintain ratios of the currents in the elements of the antenna system within a tolerance which is less than 5 percent or relative phases within tolerances which are less than three degrees. However, adjustments of transmitting equipment by such operators, except when under the immediate supervision of a radiotelephone first-class operator, are limited to the following:

(i) Those necessary to turn the transmitter on and off;

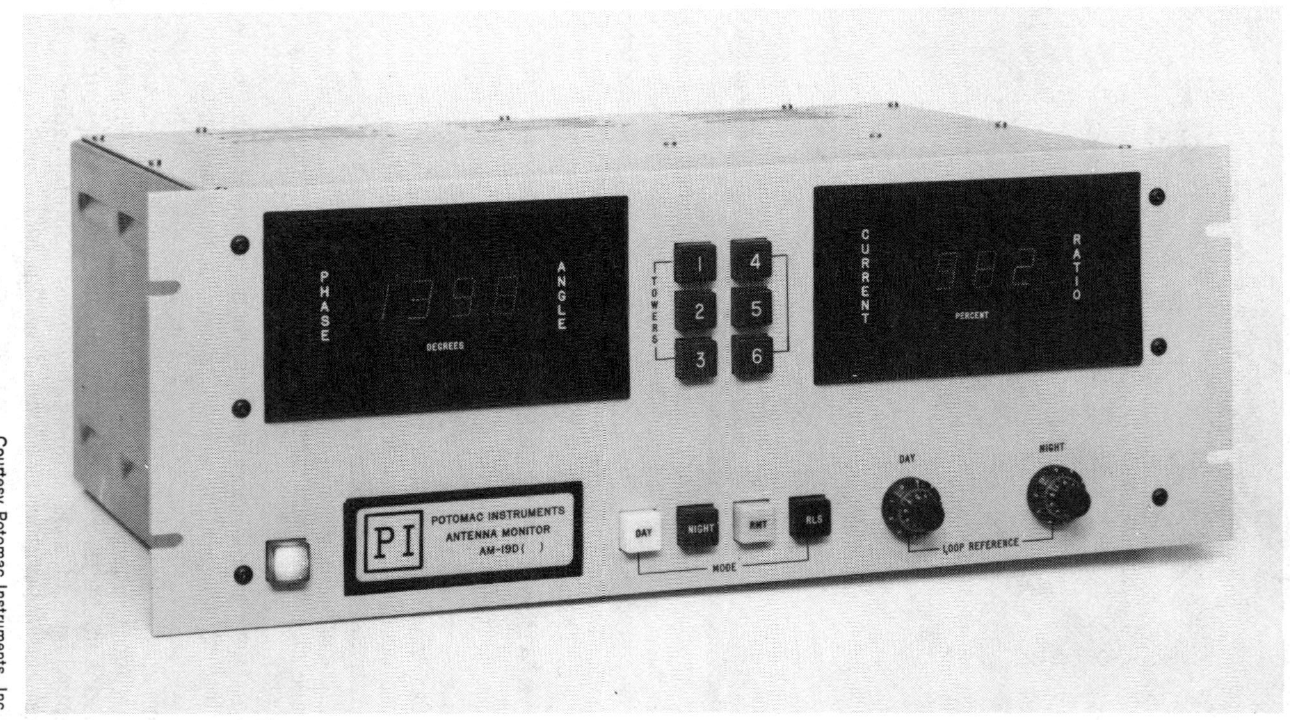

Courtesy Potomac Instruments, Inc.

Fig. 3-15. Antenna monitor.

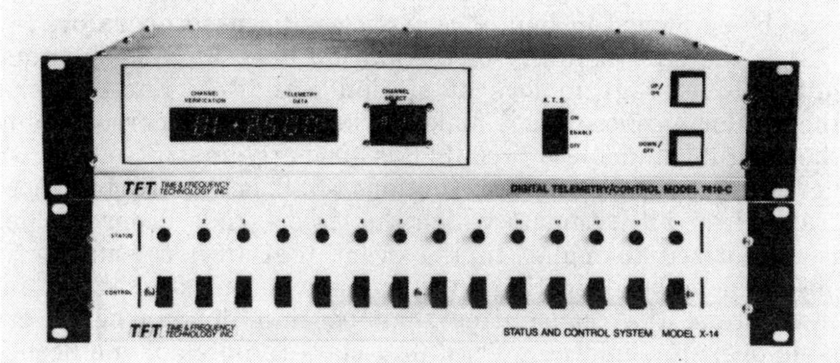

Courtesy Time and Frequency Technology, Inc.

Fig. 3-16. Digital remote-control system.

(ii) Those necessary to compensate for voltage fluctuations in the primary power supply;

(iii) Those necessary to maintain modulation levels of the transmitter within prescribed limits;

(iv) Those necessary to effect routine changes in operating power which are required by the station authorization;

(v) Those necessary to change between nondirectional and directional or between differing radiation patterns, provided that such changes require only activation of switches and do not involve the manual tuning of the transmitter's final amplifier or antenna phasor equipment. The switching equipment shall be so arranged that the failure of any relay in the directional antenna system to activate properly will cause the emission of the station to terminate.

(2) The emissions of the station shall be terminated immediately whenever the transmitting system is observed operating beyond the upper and lower limiting values of parameters required to be observed and logged or in any manner inconsistent with the rules or the station authorization, and the above adjustments are ineffective in correcting the condition of improper operation, and a first-class radiotelephone operator is not present.

(3) The special operating authority granted in this section with respect to broadcast stations is subject to the condition that there shall be in employment at the station one or more first-class radiotelephone operators authorized to make or supervise all adjustments, whose primary duty shall be to effect and insure the proper functioning of the transmitting system. In the case of a noncommercial educational fm broadcast station with authorized transmitter output power of 1000

watts or less, a second-class radiotelephone licensed operator may be employed in lieu of a first-class licensed operator.

It is the responsibility of the station licensee to keep transmitter and program logs of station activities. Furthermore, the station licensee must make certain that the person doing the logging and meter reading is properly instructed. When necessary, step-by-step instructions shall be posted for those transmitter adjustments which the lesser-grade license holder is authorized to make. In the event that the transmitter is observed to be operating in a manner inconsistent with authorization, the transmitter shall be shut down when there is no operator holding a valid first-class radiotelephone license immediately available and the authorized adjustments are not effective in correcting the condition.

LOG REQUIREMENTS

Broadcast stations are required to keep several kinds of logs. The transmitter operator on duty is required to keep the operating (transmitter) log. On occasion, he is also responsible for the program log. Frequently the program log is kept by automatic means. In some stations the operating log is also kept by an automation system. Logs must be kept by a station employee who is competent to do so, having actual knowledge of the facts. Operating logs must be signed by the person keeping the logs both at the beginning and end of his period of duty. The logs must be orderly and legible and in such detail that data required is readily available. Each log page must be numbered and dated and times shown.

Operating Log

The operating log is a technical record of the day-to-day operation of a broadcast station. Only a licensed operator may make entries in this log and they should be timely and accurate. The following rules apply to either am or fm stations:

Log the operating parameters (required transmitter and monitor meter readings) before adjustments are made to the transmitting system.

If adjustments are made, log the corrected values also. If any parameter exceeded the prescribed tolerance, add a notation describing the corrective action.

If the value of any parameter is affected by modulation, read the value without modulation. Operators at am stations frequently read the antenna current while the carrier is being modulated; the resulting reading is higher than it would have been had the carrier been unmodulated (0 percent modulation).

Specific entries required for both am and fm stations are

(1) The time each station goes on the air and the time it goes off (the time the station begins to supply power to the antenna and the time it ceases to do so).
(2) Tower lights.
   (a) If manually controlled, the time they are turned on and off.
   (b) The time of the daily check for proper operation.
   (c) Extinguishment or malfunction of any tower light.
(3) A notation concerning either activation or tests of the Emergency Broadcast System.

Additional entries required at the time each station begins operation (in each mode) and thereafter at intervals not to exceed 3 hours are:

AM
Plate voltage (E_P).
Plate current (I_P).
Antenna current or common-point current.
And, for directional am stations:
Phase.
Antenna monitor sample currents or remote antenna base current or current ratios.
FM
Plate voltage (E_P).
Plate current (I_P).
RF transmission line meter reading only if direct method used.

Furthermore, in any am station (directional or nondirectional) which determines operating power by the indirect method ($P = E_P I_P F$), the value of F (the efficiency factor) applicable to each mode of operation must be entered in the operating log once each day. Each value of F must be accompanied by a notation explaining how it was derived.

The word "mode" refers to a unique combination of transmitter, operating power, and antenna pattern. For example, if an am station is operating at 1000 watts with a nondirectional antenna, that is its mode of operation. If it reduces power to 250 watts and changes to a directional pattern at sunset, that is a different mode of operation. If it shifted from a main to an alternate main transmitter at midnight, that is a third mode, and, if at 6:00 am under a Presunrise Service

Authority, it increased power to 500 watts, that is a fourth mode of operation for the station.

MALFUNCTIONS

The recommendations of the field operations bureau of the Federal Communications Commission in case of transmitter malfunctions are as follows:

The more common malfunctions an operator is likely to encounter at a radio broadcast station are discussed. Following each example is a brief summary of the appropriate action to be taken by the operator. It is assumed that the operator has been unable to correct the problem using permitted adjustments.

The term malfunction is to be understood in the usual sense of "failure to function properly" and can be applied to either the transmitting system or the metering and monitoring equipment.

Whenever the transmitting system is observed operating:

(1) beyond the posted parameters, or
(2) in any other manner inconsistent with the rules, or
(3) in any other manner inconsistent with the terms of the station license, and, the adjustments a restricted permit holder is permitted to make cannot correct the condition, and, a broadcast technician operator is not present, then the transmitter must be turned off.

All AM and FM Stations

Modulation Monitor Fails

Continue operating; notify station licensee who will arrange for alternate means of monitoring.

Overmodulation

A serious malfunction; see above.

Overpower

A serious malfunction; see above (e.g., could happen at sunset when operator unable to reduce from daytime to nighttime power).

Underpower

Continue operating; notify station licensee (e.g., could happen at sunrise when operator unable to increase power to daytime conditions).

EBS Monitor Fails

Notify station licensee; monitor using another radio receiver if available.

Tower Lighting—Any Top Light or Any Flashing Beacon Fails

Notify station licensee; notify FAA if unable to correct within 30 minutes; make entry in operating log.

Tower Lighting—Intermediate (Side) Light Fails

Notify station licensee; make entry in operating log.

Plate Voltmeter or Plate Ammeter Fails

Notify station licensee.

AM Stations

Remote Antenna Ammeter Fails

Use regular antenna ammeter; read and log antenna current once each day for each mode of operation.

Remote and Regular Antenna Ammeters Fail

Determine operating power by the indirect method; log the product $E_P \times I_P$ or $E_P \times I_P \times F$ for each entry of E_P and I_P; log the value of F daily for each mode of operation.

FM Stations

Plate Voltmeter or Plate Ammeter Fails

Monitor power by means of rf transmission line meter.

RF Transmission Line Meter Fails

Determine power by indirect method.

Stations Operated by Remote Control

Malfunctions Resulting in Improper Control

Cease operating by remote control immediately; station may continue or resume broadcasting only if an operator takes control at transmitter.

Malfunctions Resulting in Inaccurate Meter Readings

As above except stations have one hour to correct malfunctions or post operator at transmitter or turn transmitter off.

Directional AM Stations

Remote Common-Point Ammeter Fails

Use regular common-point ammeter; read and log antenna base currents once each day for each mode of operation.

Remote and Regular Common-Point Ammeter Fail

Determine operating power by the indirect method; log the product of $E_P \times I_P$ or $E_P \times I_P \times F$ for each entry of E_P and I_P; log the value of F daily for each mode of operation.

Antenna Monitor Fails

Continue operating; notify station licensee.

Antenna Fails to Switch From Daytime to Nighttime Pattern

A serious malfunction; see above.

Antenna Fails to Switch From Nighttime to Daytime Pattern

A serious malfunction; see above.

Stations Employing Extension Meters

Malfunction of Any Component of the Extension Meter System

The pertinent entries required in the station operating log must be read and logged from the meters located at the transmitter or incorporated in the antenna monitor. If the malfunction affects indications of modulation, another acceptable means of measuring modulation must be provided at the extension meter location.

SUMMARY

A summary of station and operator responsibilities is as follows:

The Federal Communications Commission's rules and regulations derive their authority from the Communications Act of 1934.

A station license authorizes a radio station to conduct a broadcast service. It contains the call letters of the station, community, operating parameters, and other technical and legal information including special operational instructions.

Stations employing lesser-grade license holders must post instructions for those adjustments which the lesser-grade operators are permitted to make and a chart of the upper and lower limiting values of the parameters required to be observed and logged.

Each station must maintain operating, program, and maintenance logs.

Each log must be maintained by employees competent to do so and having actual knowledge of the facts.

All required entries in the operating log must be made by a licensed operator.

All required entries in the maintenance log must be made by a broadcast-technician license holder.

The operating and program logs must be signed when starting duty and again when going off duty.

No log shall be erased or obliterated.

Corrections are to be made by striking out the erroneous portion, writing in the correct entry, and initialing the correction.

The value of any parameter affected by modulation must be read without modulation.

The actual time of observation must be recorded for each log entry.

A mode of operation is a particular combination of transmitter, operating power, and antenna pattern.

If an operator:

- puts a station on the air
- changes transmitters,
- changes operating power, or
- changes antenna patterns,

he has begun a new mode of operation and must log all required operating parameters.

Thereafter, he must record a set of readings at least every 3 hours if it is not done automatically.

If at any time adjustments must be made to the transmitting apparatus, the operator must record a set of readings both before and after the adjustments are made to the transmitting apparatus.

In addition to the above, the operator must make routine entries in the operating log of the time the station goes on and off the air, the daily check of tower lights, and the monthly test of the EBS.

In an am station using the indirect method to calculate power, the operator must log the product of $E_p \times I_p$ or $E_p \times I_p \times F$ for each entry of E_p and I_p. The value of F must be entered once each day.

Definitions

Standard Broadcast (AM) Stations—A broadcast station licensed for the transmission of audio-modulated radiotelephone emissions primarily intended to be received by the general public, and operated on a frequency in the 535-kHz to 1605-kHz band.

Standard Broadcast Band—The band of frequencies extending from 535 kHz to 1605 kHz.

FM Broadcast Station—A broadcast station transmitting frequency modulated radiotelephone emissions primarily intended to be received by the general public, and operated on a channel in the 88-MHz to 108-MHz band.

FM Broadcast Band—The band of frequencies extending from 88 MHz to 108 MHz.

FM Stereophonic Broadcast—The transmission of a stereophonic program by an fm broadcast station utilizing the main channel and a stereophonic subchannel.

FM Subsidiary Communications Authorization (SCA)—An authorization granted to an fm station for the simultaneous transmission of one or more signals on assigned subcarrier frequencies within the assigned channel of the station. Special decoding equipment is required to receive program material furnished on the SCA subchannel. Such material, although broadcast related, is normally intended for reception by paying subscribers.

Daytime—That period of time between local sunrise and local sunset.

Nighttime—That period of time between local sunset and local sunrise.

Sunrise and Sunset—For each particular location and during any particular month, the times of sunrise and sunset are specified on most am broadcast station licenses. This is necessary because not all standard (am) broadcast stations are permitted to operate at night. In order to control objectionable skywave interference, stations which are permitted to operate at night are frequently required to change their modes of operation. These changes may involve the use of directional antenna systems, a reduction in operating power, or both, and normally occur at the sunrise and sunset times listed in the station license.

Broadcast Day—That period of time between local sunrise and 12 midnight local time.

Nominal Power—The power of a standard broadcast station as specified in a system of classification which includes the following values: 50 kW, 25 kW, 10 kW, 5 kW, 2.5 kW, 1 kW, 0.5 kW, and 0.25 kW.

Electrical Terms

Three common electrical terms are volts, amperes, and watts. One thousand volts is a kilovolt (kV), and 1000 watts is a kilowatt (kW). One one-thousandth (1/1000) of an ampere is a milliampere (mA). The operator should know how to use these terms interchangeably. For instance:

2000 volts is the same as 2 kilovolts
0.5 kilovolt is the same as 500 volts
200 watts is the same as 0.2 kilowatt
20 kilowatts is the same as 20,000 watts
0.5 ampere is the same as 500 milliamperes
250 milliamperes is the same as 0.25 ampere

ADDITIONAL RULES AND PROCEDURES

Remote-Control Equipment and Operation

Some broadcast stations have the main studio at one location and the transmitter and associated equipment at another location. Rather than have an operator on duty at the transmitter, the controls and metering functions of the transmitter may be at the studio or other location, and the operator on duty may be at this control point. A remote-control authorization must be obtained by the licensee from the FCC. Equipment must be installed at the control point that will permit the operator to perform all monitoring and operating functions required by the Commission's rules. If any part of the remote-control equipment, meters, or the associated control circuits has a malfunction which results in improper control or meter readings, operation of the transmitter by remote control must cease, and the station can remain on the air only with an operator on duty at the transmitter until the malfunction has been corrected.

An example of this type of malfunction could occur at an am station where the operator is unable to change from a non-directional to a directional antenna at sunset as requested by the terms of the station license. Unless there is a first-class technician present at the transmitter who can take control and make the required change, the station must cease transmitting.

If any part of the remote-control system has a malfunction that results in inaccurate meter readings, the station must cease operating by remote control within one hour. Note that the station has the same choices as above, only now it has one hour in which to take the required action.

If, for instance, an operator notices the plate current reads zero but a check of the antenna current shows the transmitter to be operating properly, the station could continue to operate by remote control for one more hour. At the end of that time the station must have posted an operator at the transmitter to take control or repair the malfunction, or the transmitter must be turned off.

Posting of Operator Permits

The operator must post his permit or posting statement at the transmitter control point where he is on duty. FCC Form 759 is a posting statement used by operators employed at more than one station. The permit is posted at one station and FCC Forms 759 are posted at other stations where the operator is employed. The station license and other operating authorizations must be posted at the transmitter control point with all terms visible.

Depending upon the operating facility of the station, the operator must post either his permit or Form 759 at the transmitter, the remote-control point, or the extension meter location.

Station Inspection and Availability of Records

The licensee of any radio station shall make the station available for inspection by representatives of the Commission at any reasonable hour. Proofs of performance, logs, measurement records, and other documents required to be maintained must also be available for inspection.

FM Stereo and Subsidiary Communications Authorizations

In addition to monaural operation, fm stations may elect to broadcast stereo programming. This is accomplished by inserting a subchannel within the main fm channel assignment. In addition to the stereophonic subchannel, an entirely different subchannel may also be used. Stations using this subchannel operate under a Subsidiary Communications Authorization (SCA). Programs transmitted by way of the SCA subchannel cannot be received without a special multiplex receiver. Most SCA subcarriers are used for the transmission of subscription background music. Other SCA uses are for detailed weather forecasting, special time signals, and other material of a broadcast nature expressly designed and intended for business, professional, educational, religious, trade, labor, agricultural, or other groups engaged in any lawful activity.

Station Identification

Broadcast-station identification announcements shall be made at the beginning and ending of each time of operation and hourly, as close to the hour as feasible, at a natural break in program offerings. Official station identification shall consist of the station call letters immediately followed by the name of the community or communities specified on the license as the station's location.

Broadcast of Taped or Recorded Material

Any taped or recorded program material in which time is of special significance, or by which an affirmative attempt is made to create the impression that it is occurring simultaneously with the broadcast, must be announced at the beginning as taped or recorded. The language of the announcement shall be clear and in terms commonly understood by the public.

Rebroadcast

The term "rebroadcast" means off-the-air reception by radio of a program originated by another radio station and its simultaneous or subsequent retransmission to the public. No broadcast station shall rebroadcast a program or part of a program of another broadcast station without the permission of the originating station. A copy of the written consent of the licensee originating the program shall be kept by the station rebroadcasting the program.

Sponsor Identification

When a broadcast station transmits matter for which it receives or expects to receive any compensation such as money, services, or other consideration, the station must announce that the matter is sponsored or furnished and also announce the name of the sponsor. However, when a commercial product or service is advertised, use of the sponsor's corporate name, trade name, or name of the product or service will suffice if it is clear that the product name identifies the sponsor. Broadcast station licensees are responsible to ensure that their employees or others connected with program material are fully aware that accurate sponsor identification must be obtained and broadcast. Sponsored programs and announcements that do not promote a specific commercial service or product must also be identified by the name of the actual sponsor. Operators keeping program logs must be alert to ensure that sponsored matter is announced at the time of the broadcast and the sponsor is correctly identified in the program log.

Lotteries

Broadcast stations are prohibited from transmitting announcements or programs promoting or containing other information that would promote lotteries. A lottery is any scheme in which money or a prize of value is awarded to a person selected by lot or chance, if a condition of winning is that a person must have furnished any money, purchased a particular product or service, or have in his possession a product sold by the sponsor. For example, a door prize given away to a person selected from tickets purchased to gain entry to a particular event is considered a lottery, and the door prize drawing cannot be promoted by radio announcements. Prizes given away to winners selected by lot when "no purchase is necessary" to enter the drawing would not be considered lottery prizes and announcements promoting these "free-entry" drawings may be broadcast. Broadcast station operators who are monitoring program material should be aware of the prohibition of broadcasting any type of lottery promotion.

Under certain conditions, advertisements and information regarding state lotteries are exempt from these prohibitions.

Broadcast of Telephone Conversations

Before broadcasting a telephone conversation or recording a telephone conversation for later broadcast, the parties to the conversation must be informed of the intention to broadcast the conversation. However, station employees who may be presumed to be aware that their telephone conversations are intended for broadcast are not required to be advised of the broadcast of their calls. Also, no notice of broadcast is required for persons originating calls to programs which normally broadcast telephone conversations coming into the program.

Emeregency Broadcast System (EBS)

The Emergency Broadcast System provides the President of the United States and the heads of state and local governments with a means of communicating with the general public in the event of a major or widespread emergency. The national EBS can be activated only on order from the President. An

Emergency Broadcast System may also be activated at the state and local levels by the appropriate officials.

An EBS check list is supplied to the station. Its purpose is to outline the correct operating procedures and announcements for every station to use in a test or real emergency. All operators should be familiar with these instructions.

All stations must install and operate, during their hours of broadcast operation, monitoring equipment capable of receiving EBS notification transmitted by other radio broadcast stations. This equipment is normally referred to as an EBS monitor receiver and need be no more than a receiver tuned to a designated station. Someone must listen to the monitor at all times during the hours of station operation. More commonly the EBS monitor is equipped with a circuit that mutes the loudspeaker. When this type of monitor receives an Attention Signal it activates an alarm and the loudspeaker, which in turn alerts the operator to take appropriate action.

Stations may or may not participate in the Emergency Broadcast System at the national, state, or local level. Note that all stations must have an EBS monitor receiver, must conduct tests, and must respond to a national-level emergency notification. The term participate is used in the sense that stations may elect to participate or not participate in the broadcasting of emergency conmmuications. Nonparticipating stations also broadcast the announcement given in the check list, but then remove their carriers from the air for the duration of the emergency.

Tower Lighting

Once each day, the operator is required to check for the proper operation of the tower lighting system. Most tower lighting systems are turned on and off with an automatic actuation switch controlled by a photocell. Stations using the photocell are required to burn the lights continuously if the automatic actuation switch does not operate properly. Any observed or otherwise known outage or improper functioning of a code or rotating beacon light or top light not corrected in 30 minutes should be reported by telephone or telegraph to the nearest Flight Service Station or office of the FAA. Notice should also be given to the Flight Service Station or FAA office when the situation has been corrected.

Duplicate and Renewed Commercial Operator Permits

If an operator's license or permit is lost or destroyed, a duplicate license or permit may be obtained by application to the FCC office where the lost or destroyed document was issued. As long as the license or permit has not expired, the operator may continue to operate the station. If the operator's license is required to be posted at his place of duty, a copy of the application, FCC Form 756, submitted for the duplicate license may be posted in lieu of the license pending receipt of the duplicate license.

Operator licenses may be renewed any time during the final year of the license term or within a one-year grace period after the expiration of the license by submitting FCC Form 756 to the nearest FCC field office. If the application for renewal is submitted before the license expires, the operator may continue operating unless he hears otherwise from the Commission. A duplicate copy of the renewal application should be posted in lieu of the license, since the license being renewed must accompany the application.

If the renewal application is submitted during the grace period, the license may still be renewed, but the operator has no operating authority between the expiration of the previous license and the issuance of the renewed license.

CHAPTER 4

Broadcast Study Guide and Self-Examination

Although no examination is required to obtain the restricted operator permit, it is helpful to gain as much knowledge as possible about broadcast operations and the associated FCC Rules and Regulations. In fact, the questions that follow emphasize those factors that the FCC considers important in the proper operation of a broadcast station. It is this information that will become part of your instructions when you are employed by a broadcast station and assume the responsibilities assigned to you by the first-class licensed technician at the station.

It is important that a person preparing for employment at a broadcast station know the material contained in Chapter 3. Even though the licensee holds only a restricted radiotelephone operator permit he should be familiar with operating procedures, some transmitter fundamentals, and have a knowledge of the appropriate FCC Rules and Regulations.

This chapter serves as a review and study guide. One always has a better awareness of knowledge gained when questioned directly. The first set of questions are based on a sample testing sequence prepared by the Field Operations Bureau of the FCC. Additional questions are given on the material in Chapter 3 along with appropriate answers. This chapter concludes with a self-test that will be of further assistance in consolidating the information given in Chapter 3.

MULTIPLE-CHOICE TEST

1. When the remote-control equipment malfunctions and results in improper transmitter control, the
   A. operation by remote control must cease.
   B. power must be reduced by 5 percent.
   C. power must be reduced by 10 percent.
   D. station must be taken off the air.
   E. FCC must be notified.

2. The antenna-current meter, or remote antenna-current meter where provided, should be read
   A. during modulation peaks.
   B. prior to placing the station on the air.
   C. when the program material modulates the carrier at the average level.
   D. when the carrier is unmodulated.
   E. once each hour.

3. The PRIMARY concern of an operator on duty at an fm broadcast station should be to
   A. ensure that the transmitter is operating within required parameters.
   B. ensure "spot" announcements are broadcast and logged properly in the program log.
   C. maintain a "tight board."
   D. keep the station on the air at all cost.
   E. keep the chief operator advised of problems encountered during the shift.

4. A standard (am) broadcast station determines the operating power by the direct method. The meter indications that must be routinely observed and entered into the operating log are
   A. plate voltage, plate current, and rf transmission line.
   B. plate voltage, plate current, filament voltage, and modulation.
   C. plate voltage, plate current, and antenna or common-point current.
   D. plate voltage, plate current, and peak modulation.
   E. plate voltage and current only.

5. The required broadcast station identification announcement given during the broadcast day must include the
   A. licensee's name and location.
   B. same identification as required during sign-off.
   C. assigned frequency and call sign.
   D. call sign and station affiliation.
   E. assigned frequency and power.

6. Which of the following are not required to be recorded in the operating log?
   A. Antenna ammeter reading.
   B. Plate-voltage meter reading.
   C. Plate-current meter reading.
   D. Modulation monitor meter reading.
   E. The time the station ceases to supply power to the antenna.

7. In the event the modulation monitor becomes defective, an fm broad-
cast station may be operated without the monitor provided that
   A. the audio control is adjusted to midrange.
   B. the percentage of modulation is monitored with an oscilloscope or
   other acceptable means.
   C. the transmitter power is reduced to half of its licensed power.
   D. the operator makes daily entries in the operating log.
   E. a first-class operator is on duty.

8. A lower-grade operator on duty at a broadcast station is authorized to
   A. adjust external controls to compensate for voltage fluctuations in
   the primary power supply.
   B. make transmitter adjustments to maintain the correct operating
   frequency.
   C. repair an inoperative transmitter in an emergency.
   D. replace defective final amplifier tubes.
   E. make any necessary minor adjustments of internal transmitter
   controls.

9. If both the antenna ammeter and remote antenna ammeter of an am
broadcast station become defective, the operating power must then
be determined by the
   A. rf-transmission-line meter reading method.
   B. power-meter reading method.
   C. indirect method.
   D. final amplifier method.
   E. percentage variance method.

10. If a flashing beacon at some intermediate level on an antenna tower
burns out, the
   A. operator should notify the local FCC field office.
   B. operator should notify the nearest FAA office or flight service sta-
   tion if it cannot be restored within 30 minutes.
   C. other lights should be left burning continuously.
   D. other lights should be turned off until it is repaired.
   E. local power company should be notified.

11. The operating log must contain
   A. the name of all sponsors of a sponsored program.
   B. the time of all recorded program announcements.
   C. an entry of the time the transmitter begins to supply power to
   the antenna.
   D. the time station identification announcements were given.
   E. notations concerning maintenance of the transmitter.

12. The Commission's rules specify that the operating power shall not ex-
ceed the limit of
   A. 125 percent of authorized power.
   B. 120 percent of authorized power.
   C. 115 percent of authorized power.
   D. 110 percent of authorized power.
   E. 105 percent of authorized power.

13. If an automatic alarm system is not in service, observations to ensure
that the tower lights at a station are functioning properly must be
made at least once every

A. three months.
B. week.
C. twenty-four hours.
D. twelve hours
E. hour between sunset and sunrise.

14. Entries in the am or fm station operating log shall be made by
    A. any station employee having knowledge of the required facts.
    B. the properly licensed operator in actual charge of the transmitter.
    C. any employee duly authorized to do so by the station licensee.
    D. the full time or contract maintenance operator.
    E. any licensed radio operator.

15. If the modulation-monitor peak indicator (flasher) in a standard broad-
    cast station indicates peaks of frequent recurrence in excess of 100
    percent, what action should be taken by the operator?
    A. Adjust filament voltage to normal.
    B. Inform the chief operator.
    C. Reduce the transmitter final plate voltage.
    D. Reduce the audio level.
    E. Notify the station manager.

16. An operator holding the proper grade license or permit shall be in ac-
    tual charge of the transmitting apparatus and shall be on duty at
    A. any location on the station premises so long as he can make the
       required meter readings at the specified intervals and the equip-
       ment is provided with failure alarms.
    B. any location on the station premises if programming and trans-
       mitter logging is automated.
    C. any location on the station premises if programming is automated.
    D. any location on the station premises where the required metering
       and control equipment is installed.

17. Operating power at an am broadcast station, when determined by the
    direct method, is computed or obtained from
    A. the sum of the antenna current plus the plate current.
    B. the square of the antenna current times the antenna resistance.
    C. the antenna current times the square of the antenna resistance.
    D. the plate voltage times the plate current.
    E. the calibrated power meter.

18. An operator at a broadcast station not operated by remote control must
    post his operator license or permit
    A. at any location at the station where it is readily visible.
    B. where it can be inspected by the public.
    C. at any place where it is available for inspection by FCC repre-
       sentatives.
    D. in the main studio.
    E. at the transmitter or extension meter location.

19. An am broadcast station is not permitted to modulate the carrier to
    A. 0 percent
    B. 50 percent on negative peaks of frequent recurrence.
    C. 85 percent on positive peaks of frequent recurrence.
    D. 105 percent on negative peaks of frequent recurrence.
    E. 120 percent on positive peaks of frequent recurrence.

20. A radio broadcast station operator discovers that his operator permit
has expired two weeks previously. The operator may not continue to
operate the transmitter
    A. until the station manager notifies the FCC.
    B. until a renewed permit has been issued.
    C. until he has been re-examined.
    D. unless an application for renewal is filed immediately and a copy
       posted.
    E. unless certified to do so by a first-class operator.

21. Station identification must be given
    A. on the hour.
    B. on the half hour.
    C. at 30 minute intervals.
    D. after a commercial.
    E. after each newscast.

22. The plate voltmeter reads 2 kilovolts. This is the same as
    A. 0.002 volt.
    B. 2 volts.
    C. 20 volts
    D. 200 volts.
    E. 2000 volts.

23. The plate voltmeter at an fm broadcast station reads 2 kilovolts, and
the plate current meter reads 500 milliamperes. With an efficiency of
70 percent, the operating power as determined by the indirect method is
    A. 140 watts.
    B. 350 watts.
    C. 700 watts.
    D. 1000 watts.
    E. 1400 watts.

24. The antenna resistance at a standard broadcast station is 40 ohms, and
the antenna current is 5 amperes. The power calculated by the direct
method is
    A. 8 watts
    B. 200 watts.
    C. 900 watts.
    D. 1000 watts.
    E. 8000 watts.

25. Tower one of a three-tower directional antenna array is the reference
tower. The antenna current for tower one is 6 amperes. The antenna
current for tower two is 2 amperes, and the antenna current for tower
three is 3 amperes. The antenna base-current ratio of tower two is
    A. 0.33
    B. 0.5
    C. 1
    D. 2
    E. 3

1. A	8. A	14. B	20. B
2. D	9. C	15. D	21. A
3. A	10. B	16. D	22. E
4. C	11. C	17. B	23. C
5. B	12. E	18. E	24. D
6. D	13. C	19. D	25. A
7. B			

STUDY QUESTIONS

The following study questions, which require a direct answer, emphasize important FCC Rules and Regulations. Many of these topics were covered in Chapter 3 and serve as an excellent review.

1. **What is meant by the following words or phrases: Standard broadcast station, standard broadcast band, standard broadcast channel, fm station, fm band, daytime, nighttime, broadcast day, and EBS?**—Refer to appropriate definitions in Chapter 3.

2. **Make the following transformations: Kilohertz to hertz, kilovolts to volts, milliamperes to amperes.**—To convert kilohertz to hertz it is necessary to multiply the number of kilohertz by 1000. To convert kilovolts to volts it is necessary to multiply the number of kilovolts by 1000. To convert milliamperes to amperes it is necessary to divide the number of milliamperes by 1000.

3. **Draw the faces of the following meters and know how to read each: plate current and plate voltage, antenna current and radio-frequency kilowatts, modulation percentage, and antenna phasing in degrees.**

4. **What should an operator do if the remote antenna meter becomes defective?**—A remote antenna ammeter is not required; therefore, authority to operate without it is not necessary if it becomes defective. However, if the remote antenna ammeter does becomes defective, the antenna base current may be read and logged once daily for each mode of operation, pending the return to service of the regular remote antenna ammeter.

5. **What should an operator do if the remote-control devices at a station so equipped malfunction?**—It shall be cause for the immediate cessation of remote-control operation.

6. **What is the permissible percentage of modulation for am and fm stations?**—The percentage of modulation shall be

maintained as high as possible consistent with good quality of transmission. Generally, it should be no less than 85 percent on peaks of frequent recurrence, except as necessary to avoid objectionable loudness. For am, it shall not exceed 105 percent on negative peaks of frequent recurrence or 125 percent on any positive peaks. For fm, it shall not exceed 100 percent on peaks of frequent recurrence.

7. **What is the permissible frequency tolerance of standard am broadcast stations? Of fm stations?**—The operating frequency of the am broadcast station shall be maintained within 20 Hz of the assigned frequency. The center frequency of each fm broadcast station shall be maintained within 2000 Hz of the assigned center frequency.

8. **What are the power limitations on broadcast stations?** —The operating power of each station will be maintained as near as practicable to the licensed power and shall not exceed the limits of 5 percent above and 10 percent below the licensed value, except in an emergency when, due to causes beyond the control of the licensee, it becomes impossible to operate with full power.

9. **What logs must be kept by broadcast stations, according to the Rules and Regulations of the FCC?**—Program, operating, and maintenance logs.

10. **Who keeps the logs?**—They must be kept by person or persons competent to do so and having knowledge of the facts required, and by the operator on duty.

11. **When may abbreviations be used in the station's logs?**— Only when proper meaning or explanation is contained elsewhere in the log.

12. **How and by whom may station logs be corrected?**—If a correction is made before the person keeping the log has signed the log upon going off duty, such correction, no matter by whom made, shall be initialed by the person keeping the log prior to his signing of the log when going off duty, as attesting to the fact that the log as corrected is an accurate representation of what was broadcast. If corrections or additions are made on the log after it has been so signed, explanation must be made on the log or an attachment to it, dated and signed by the person who kept the log. Any necessary correction shall be made by striking out the erroneous portion or by making a corrective explanation on the log or attachment to it.

13. **According to the Rules and Regulations of the FCC, how long must a station's logs be retained?**—For a period of two years.

14. **What information must be given an FCC inspector at any reasonable hour?**—Program, operating, and maintenance logs, equipment performance measurements, a copy of the most recent antenna resistance or common-point impedance measurement submitted to the commission, and a copy of the most recent field-intensity measurement to establish performance of the directional antenna.
15. **What is included in a station identification and how often is it given?**—Call letters and location. Station identification shall be made (1) at the beginning and ending of each time of operation and (2) hourly, as close to the hour as feasible, at a natural break in program offerings.
16. **What should an operator do if the modulation monitor becomes defective?**—The station may be operated without the monitor pending its repair or replacement for a period not in excess of 60 days provided appropriate maintenance log entries are made, and the degree of modulation is monitored with an oscilloscope or other acceptable means.
17. **When should minor corrections to the transmitter be made, before or after logging the meter reading?**—Logging should be done prior to making any adjustment to the equipment. When appropriate, an indication of the correction made should be noted.
18. **How often should the tower lights be checked for proper operation?**—Once each 24 hours.
19. **What record is kept of tower-light operation?**—On and off times for each ray, time of daily check of proper operation, any observed or otherwise-known failure of a tower light, and results of quarterly inspections.
20. **What should an operator do if the tower lights fail?**—The failure is to be reported immediately by telephone or telegraph to the nearest Flight Service Station or office of the FAA if not corrected within 30 minutes.
21. **What equipment must be installed in broadcast stations in regard to the reception of an emergency-action notification?**—Necessary equipment to receive emergency-action notification or termination by specified means. This equipment must be maintained in a state of readiness for reception, including arrangements for operator listening watch or automatic alarm devices.
22. **If tower lights of a station are required to be controlled by a light-sensitive device and this device malfunctions, when should the tower lights be on?**—They shall burn continuously.

SELF-TEST

1. One million hertz is how many kilohertz?

A. One million.
B. One thousand.
C. One hundred.
D. Ten thousand.

2. The remote antenna meter becomes defective. You must

A. call FCC.
B. shut down transmitter.
C. continue operation.
D. notify FAA.

3. What deviation constitutes 100 percent modulation for an fm broadcast transmitter?

A. ±75 kHz.
B. ±25 kHz.
C. 2000 Hz.
D. 20 Hz.

4. Your modulation meter reads much above 100 percent on peaks. You must

A. increase plate voltage.
B. decrease plate current.
C. reduce volume levels.
D. readjust modulation meter.

5. A restricted permit operator may operate

A. a directional am station with antenna current-ratio tolerance less than 5 percent.
B. a radiotelegraph transmitter
C. a television station.
D. none of the above.

6. A broadcast station must keep

A. only one log.
B. emergency log only.
C. no program log.
D. program and operating logs.

7. An operating log must include

A. time of sign-on and sign-off.
B. final stage current readings.
C. service interruptions.
D. all of above.

8. A program log must not include

A. name of sponsor.
B. identification times.
C. name of announcer.
D. time of spot commercial.

9. Station logs may be corrected by the

A. general manager.
B. chief engineer.
C. FCC inspector.
D. person making mistake.

10. FCC inspector must be given upon request

A. any station log.
B. income report for the week.
C. names of all employees and their incomes.
D. all of above.

11. **If a modulation monitor fails**

A. the station must shut down.
B. notify the first-class operator.
C. forget about keeping the log.
D. turn up the volume level.

12. **Scheduled operating logging should be made**

A. each hour only.
B. after corrective adjustments.
C. prior to corrective adjustments.
D. at the end of the day only.

13. **An announcement of prerecorded material must be made**

A. at sign-off time.
B. on the half-hour.
C. to avoid any false impression that program is live.
D. only on network shows.

14. **What is considered a broadcast day?**

A. Local sunrise to midnight local time.
B. Local sunrise to sunset.
C. 6:00 am to 8:00 pm.
D. Daytime.

15. **2000 volts is how many kilovolts.**

A. 2000.
B. 2.
C. 20.
D. 1/2000.

16. **Frequency of the fm band is**

A. 540 to 1600 kHz.
B. 88 to 108 MHz.
C. 2000 Hz.
D. ±75 kHz.

17. **If the reading on Meter 2 in Fig. 3-1 dropped 10 percent, what would it read?**

A. 3620.
B. 2880.
C. 2000.
D. 350.

18. **With overmodulation, Meter 5 in Fig. 3-1 would read**

A. 89 percent.
B. 100 percent.
C. above 100 percent.
D. under 85 percent.

19. **If the reading of Meter 3 in Fig. 3-1 increased 5 percent, it would read**

A. 1.7.
B. 1.9.
C. 2.70.
D. 1.785.

20. **If the remote-control system fails, you should**

A. shut down the station.
B. call the chief engineer.
C. call the FCC inspector.
D. call the telephone company.

21. **A frequency monitor**

A. measures carrier drift.
B. need not be part of remote control monitoring.
C. is not a compulsory test unit.
D. all of the above.

22. **Frequency tolerance for an am station is**

A. 20 Hz.
B. 2000 Hz.
C. 0.01 percent.
D. 0.0001 percent.

23. **Frequency tolerance for an fm station is**

A. 0.0001 percent.
B. 0.001 percent.
C. 2000 Hz.
D. 20 Hz.

24. **What are the maximum and minimum power limits for a 10-kW am station?**

A. 11 and 9 kW.
B. 10.5 and 9.5 kW.
C. 11 and 9.5 kW.
D. 10.5 and 9 kW.

25. **Who always keeps radio station logs?**

A. A person competent to do so and with knowledge of facts.
B. A chief engineer.
C. A general manager.
D. Operator on duty.

26. **May abbreviations be used in logs?**

A. Never.
B. Anytime.
C. According to log list.
D. After approval by FCC inspector.

27. **Logs must be retained for**

A. one year.
B. two years.
C. five years.
D. indefinitely.

28. **Call letters and location must be given**

A. at sign-on.
B. at sign-off.
C. on the half-hour.
D. A and B.

29. **If frequency meter fails**

A. station must be shut down.
B. notify first-class operator.
C. forget about logging it.
D. turn up power output.

30. **When can sponsor's name be omitted from a commercial?**

A. Never.
B. Upon his request.
C. When station manager approves.
D. During nonprime time.

31. **Tower lights must be checked**

A. daily.
B. weekly.
C. monthly.
D. hourly.

32. **What is the procedure to follow if tower lights fail?**

A. Shut down station.
B. Call FCC immediately.
C. Call FAA if repair is not possible in one-half hour.
D. Wait for morning.

33. **What record must be kept of tower-light operation?**

A. On and off times.
B. Failures.
C. Daily inspection.
D. All of above.

34. **What is an EBS receiver?**

35. **During a period of emergency-action condition what should nonparticipating stations do?**

36. **What is the modulation percentage indicated by meter A in Fig. 4-1?**

A. 90.
B. −2.
C. 100.
D. 95.

37. **What does meter B in Fig. 4-1 suggest?**

A. Your fm transmitter is off frequency.
B. Everything is just fine with am transmitter.
C. The am transmitter is 20 Hz low.
D. Temperature is too low.

38. **Arbitrary meter C in Fig. 4-1 is reading**

A. 70.
B. 70+.
C. high.
D. low.

39. **Voltmeter D in Fig. 4-1 is reading**

A. 1550 volts.
B. too high.
C. 1500 volts.
D. 1450 volts.

40. **Ammeter (Fig. 4-1) is reading**

A. 325 milliamperes.
B. too high.
C. 350 milliamperes.
D. 350 amperes.

41. **Indirect power to an antenna is**

A. $I_A^2 R_A$
B. $I_P^2 E_P$
C. measured with a field intensity meter.
D. a substitute method of transmitter power measurement.

42. **The direct power to an antenna is**

A. $I_A^2 R_A$
B. $I_P^2 E_P$
C. measured with a field intensity meter.
D. taken with a plate current meter.

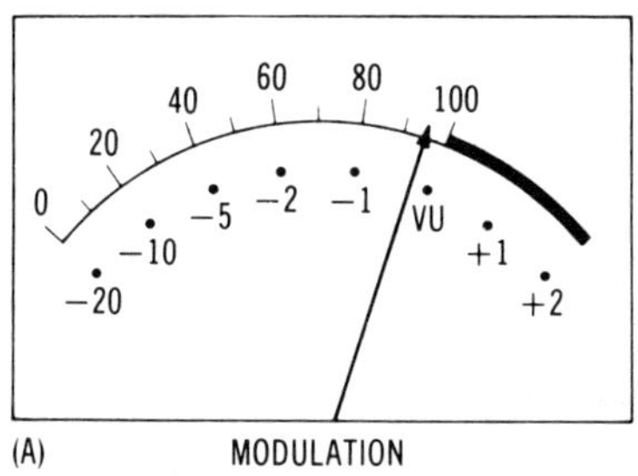

(A)   MODULATION

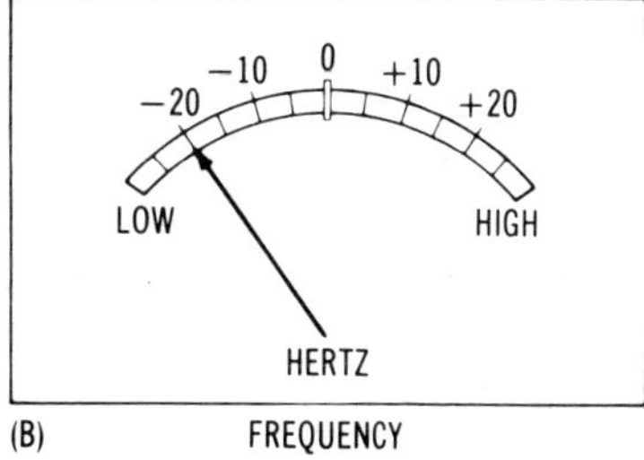

(B)   FREQUENCY

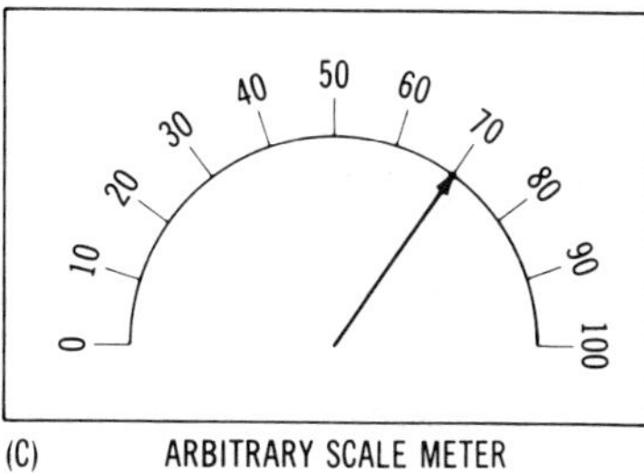

(C)   ARBITRARY SCALE METER

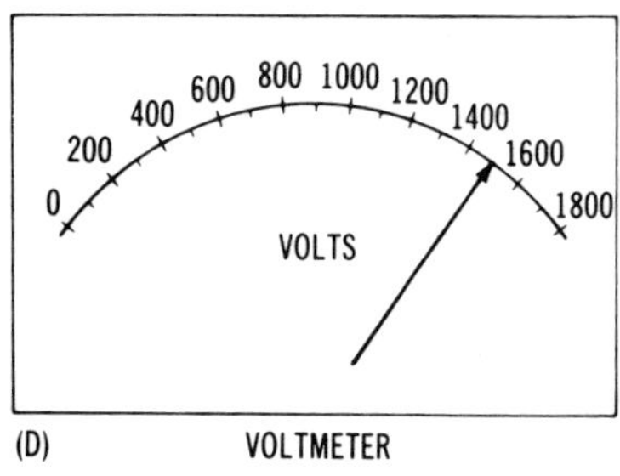

(D)   VOLTMETER

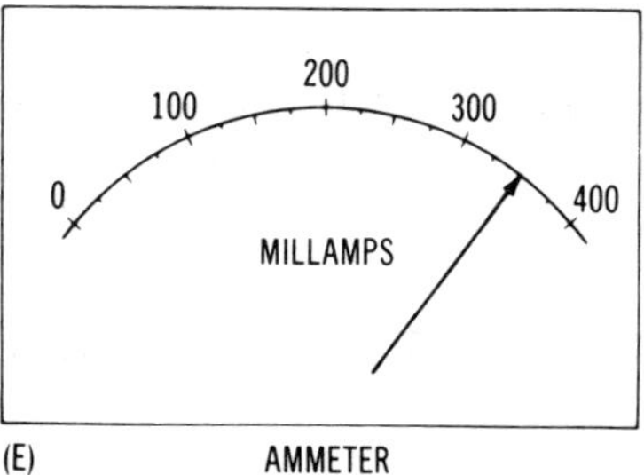

(E)   AMMETER

Fig. 4-1. Meters for questions 36 through 40.

43. The antenna current of a standard broadcast station

A. is measured in the plate circuit.

B. is measured with no modulation.

C. is measured with dummy load.

D. does not change with modulation.

44. The output power for an fm station is often measured

A. by indirect means.

B. by direct means with no output meter.

C. at top of the tower.

D. once each week.

45. Give the definition of antenna current.

46. What does antenna resistance mean?

47. Give the definition of antenna power.

48. What is a rebroadcast?

49. Operation by indirect power measurement is permissible when

A. modulation fails.
B. the plate-circuit-current meter fails.

C. switching to directional antenna.
D. antenna-current meter fails.

50. What is transmitter power input if plate voltage is 3500 volts and dc plate current is 285 milliamperes?

A. 99.75 watts.
B. 12.3 kilowatts.

C. 1 kilowatt.
D. 9.975 kilowatts.

51. What is the power input on the basis of the readings shown by meters D and E in Fig. 4-1?

A. 5250 watts.
B. 428 watts.

C. 525 watts.
D. 4280 watts.

52. How many amperes is 300 milliamperes?

A. 0.3
B. 3.3

C. 3
D. 0.333.

53. How many kilowatts is 2000 watts?

A. 0.2
B. 2.0

C. 20
D. 5

54. The current to Tower 2 is 1.5 amperes. What is the current ratio for Tower 2 if the current at the reference tower is 2.5 amperes?

A. 0.166
B. 0.6

C. 1.66
D. 0.375

55. Both fm stereo and SCA transmissions

A. need not be logged.
B. use directional antennas.

C. are sent out on subchannels.
D. require FCC form 759.

56. Station identification must be made

A. exactly on the hour.
B. at end of operations.

C. after each commercial.
D. each half hour.

57. To rebroadcast a program of another station

A. obtain a verbal consent.
B. apply for FCC permission.

C. notify FCC local office.
D. retain a written consent.

58. The restricted permit operator at a broadcast station

A. must be able to tune a crystal oscillator.

B. conducts visitor tours.

C. must be able to operate EBS receiver.

D. keeps the transmitter keys.

59. Program logs contain

A. identification of each program.

B. identification of each commercial.

C. time devoted to commercials each hour.

D. all of above.

ANSWERS TO SELF-TEST

1. B. One kilohertz equals one thousand hertz. In one million hertz there are one thousand kilohertz.
2. C. The remote antenna meter reading is not a compulsory one and operations can continue. However, you should always notify the first-class operator.
3. A.
4. C. Too high an audio level can cause overmodulation.
5. D.
6. D.
7. D.
8. C.
9. D. You should strike out the erroneous portion (not erase) and initial the correction made and the date of correction.
10. A.
11. B. In this case there is no need to take the station off the air.
12. C. Minor readjustments are always made after the official logging.
13. C.
14. A.
15. B. One kilovolt equals one thousand volts; therefore, two thousand volts is the same as two kilovolts.
16. B.
17. B. 10 percent of the 3200-volt reading is 320 volts. Reading would be 2880 volts (3200 − 320).
18. C.
19. D. The present reading is 1.7, and 5 percent of this figure is 0.085. This would be an increase to 1.785.
20. A. In fact, failure of the remote-control system would usually shut down the transmitter automatically.
21. D.
22. A.
23. C.
24. D. The limits are 5 percent above assigned power and 10 percent below assigned power.
25. D.
26. C. Abbreviations may be used if a list of such abbreviations is kept at some point in the log.
27. B.
28. D.
29. B. This is not a condition for shutting down the station. However, it is your responsibility to immediately notify the chief engineer.
30. A.
31. A.
32. C.
33. D.
34. A required receiver that each station must have tuned to the Emergency Broadcast System station of the area.
35. All nonparticipating broadcast stations will observe broadcast silence.
36. D.
37. C.

38. A.
39. C.
40. C.
41. D.
42. A.
43. B.
44. A.
45. Antenna current means the radio-frequency current in the antenna with no modulation.
46. Antenna resistance means the total resistance of the transmitter antenna system at the operating frequency and at the point at which the antenna current is measured.
47. Antenna input power or antenna power means the product of the antenna current squared times the antenna resistance at the point where the current is measured.
48. The term rebroadcast means reception by radio of the program of a radio station, and the simultaneous or subsequent retransmission of such program by another broadcast station.
49. D.
50. C.
51. C. (1500 × 0.35.)
52. A.
53. B.
54. B. (1.5/2.5.)
55. C.
56. B.
57. D.
58. C.
59. D.

CHAPTER 5

Third-Class Radiotelephone License

The third-class radiotelephone is an important license if you are interested in marine radiotelephone operations. This license or a higher grade license is required for certain cargo and passenger ships. The third-class radiotelephone license is also a beginning license in advancing to one of the higher grade licenses.

Each cargo ship of the United States which, in accordance to Part II of Title III of the Communications Act, is equipped with a radiotelephone station shall for safety purposes carry at least one qualified operator. Where the power of the station does not exceed 250 watts carrier power or 1500 watts peak envelope power, such operator shall hold a radiotelephone third-class operator permit or higher class of operator authorization. This same requirement applies for each vessel of the United States transporting more than six passengers for hire.

The third-class radiotelephone license is obtained by passing an appropriate FCC multiple-choice examination involving Elements I and II. A study guide of questions and answers are included with this chapter. In addition there is a multiple-choice self-test examination at the end of the chapter so that you might gain experience in taking this type of test. Also study Appendices A, B, and C thoroughly.

ELEMENT I—BASIC LAW

1. **Where and how is an operator license or permit obtained?**
—Request an application form and examination schedule from the nearest FCC field office. Submit the application in the prescribed form, including all subsidiary forms and documents, properly completed and signed, in person or by mail, to the office of your choice (usually the closest one). This office will make the final arrangements. (R.R. 13.11a.)
2. **When a licensee qualifies for a higher grade of FCC license or permit, what happens to the lesser grade license?** —The license or permit held will be cancelled upon issuance of the new license. (R.R. 13.26.)
3. **Who may apply for an FCC license?**—Normally, commercial licenses are issued only to citizens of the United States. (R.R. 13.5a.)
4. **If a license or permit is lost, what action must be taken by the operator?**—The commission must be notified immediately and a properly executed application for a duplicate should be submitted. A statement must be included regarding the circumstances involved in the loss of the license. The operator must exhibit in lieu of the original document a signed copy of the submitted application. (R.R. 13.71 and 13.72.)
5. **What is the usual license term for radio operators?**— Five years. (R.R. 13.4a.)
6. **What government agency inspects radio stations in the United States?**—The Federal Communications Commission. (Sec. 303n.)
7. **When may a license be renewed?**—The application may be filed at any time during the final year of the license term or during a one-year period after the date of expiration of the license. During this one-year grace period any expired license is not valid. (R.R. 13.11a.)
8. **Who keeps the station log?**—Person or persons competent to do so having actual knowledge of the facts required. They shall sign the appropriate log when starting duty and again when going off duty. (R.R. 73.1800.)
9. **Who corrects errors in the station log?**—Only the person originating the entry. (R.R. 73.1800, 73.1810, and 73.1820.)
10. **How may errors in the station log be corrected?**—The person originating the entry shall strike out the erroneous portion, initial the correction made and indicate the date of correction. (R.R. 3.1800.)
11. **Under what conditions may messages be rebroadcast?**—

Only by the express authority of the originating station. (Sec. 325a.)

12. **What messages and signals may not be transmitted?**— Unnecessary, unidentified, or superfluous communications; obscenity, indecency, or profanity; false signals or any call or signal which has not been assigned by proper authority to the radio station concerned. (R.R. 13.66, 13.67, and 13.68.)

13. **May an operator deliberately interfere with any communication or signal?**—No. (R.R. 13.69.)

14. **What type of communication has top priority in the mobile service?**—Distress calls, distress messages, and distress traffic. (ART. 37.)

15. **What are the grounds for suspension of operator's license?** —Violation of any provision of any act, treaty, or convention binding on the United States; failure to carry out a lawful order of the master or person lawfully in charge of ship or aircraft on which he is employed; damaging or permitting the damage of any radio apparatus or installation; transmission of profane or obscene language, false or deceptive signals, or call signals or letters not assigned to the station in operation; wilful or malicious interference with other radiocommunications or frequencies; obtaining, attempting to obtain, or assisting another to obtain or attempt to obtain an operator's license by fraudulent means. (SEC. 303m.)

16. **When may an operator divulge the contents of an intercepted message?**—He may divulge the contents of any radiocommunications broadcast or transmission by others for use of the general public or relating to ships in distress. (SEC. 605.)

17. **If a licensee is notified that he has violated an FCC rule or provision of the Communications Act of 1934, what must he do?**—He must reply within ten days to the office of the commission originating the official notice. The answer to each notice shall be complete in itself and shall not be abbreviated by reference to other communications or other notices. In every instance the answer shall contain a statement of the action taken to correct the condition or omission complained of, and to preclude its recurrence. Information concerning equipment of concern, and name and license number of operator in charge. (R.R. 1.89.)

18. **If a licensee receives a notice of suspension of his license, what must he do?**—No order of suspension of any operator's license shall take effect until 15 days after a notice in

writing of the cause has been given to the operator. He may make written application to the commission at any time within said 15 days for a hearing upon such order. If, after a hearing, a license is ordered suspended, the operator shall send his license to the office of the Commission in Washington, D.C., on or before the effective date of such order. (R.R. 1.85.)

19. **What are the penalties provided for violation of the provisions of the Communications Act of 1934 or a rule of the FCC?**—Penalties for violation of the Act provide for a fine of not more than $10,000.00 or imprisonment for a term not exceeding two years, or both. The penalty provided for violation of FCC rules is a fine of not more than five hundred dollars ($500) for every day during which said offense occurs. (SEC. 501 and 502.)

20. **Define harmful interference.**—Any emission, radiation, or induction which endangers the functioning of a radio-navigation service or other safety services, or seriously degrades, obstructs, or repeatedly interrupts a radio-communication service operating in accordance with the regulations.

ELEMENT II—BASIC OPERATING PRACTICE

Each radiotelephone operator should know and abide by the rules and accepted procedures of operation. Courtesy and consideration are two very important factors in minimizing unnecessary interference and maintaining reliable communications on active channels. The radiotelephone license holder should be an example to operators of lower grade, and he should do everything possible to encourage proper operations in the service or services with which he is concerned.

A fine summary of radiotelephone operating practice, recommended by the FCC, follows:

A licensed radio operator should remember that the station he desires to operate should be licensed by the Federal Communications Commission. In order to prevent interference and to give others an opportunity to use the airways, he should avoid unnecessary calls and communications by radio. He should remember that radio signals normally travel outward from the transmitting station in many directions, and therefore these transmitted signals could be intercepted by unauthorized persons.

Before making a radio call, the operator should listen on the communications channel to ensure that interference will not be caused to communications which may already be in progress. At all times in radiocommunications the operator should be courteous.

Station identification should be made clearly and distinctly so as to avoid unnecessary repetition of call letters and enable other stations to clearly identify all calls.

An operator normally exhibits his authority to operate a station by posting a valid operator license or permit at the transmitter control point.

While a radio transmitter is in a public place, it should at all times be either attended by or supervised by a licensed operator; or the transmitter should be made inaccessible to unauthorized persons.

A radio transmitter should not be on the air except when signals are being transmitted. The operator of a radiotelephone station should not press the push-to-talk button, except when he intends to speak into the microphone. Radiation from a transmitter may cause interference even when voice is not transmitted.

When radiocommunications at a station are unreliable or disrupted due to static or fading, it is not a good practice for the operator to continuously call other stations in attempting to make contact, because his calls may cause interference to other stations that are not experiencing static or fading.

A radiotelephone operator should make an effort to train his voice for most effective radiocommunication. His voice should be loud enough to be distinctly heard by the receiving operator, and it should not be too loud, since it may become distorted and difficult to understand at the receiving station. He should articulate his words and avoid speaking in a monotone as much as possible. The working range of the transmitter is affected to some extent by the loudness of the speaker's voice. If the voice is too low, the maximum range of the transmitter cannot be attained; if the voice is too loud, the range may be reduced to zero due to the signals becoming distorted beyond intelligibility. In noisy locations the operator sometimes cups his hands over the microphone to exclude extraneous noise. Normally, the microphone is held from 2 to 6 inches from the operator's lips.

It is important in radiotelephone communications that operators use familiar and well-known words and phrases in order to ensure accuracy and eliminate undue repetition of words. Some radio operating companies, services, networks, associations, etc., select and adopt standard procedure words and phrases for expediting and clarifying radiotelephone conversations. For example in some services, "Roger" means "I have received all of your last transmission." "Wilco" means "Your last message received, understood, and will be complied with," "Out" or "Clear" means "This conversation is ended and no response is expected," "Over" means "My transmission is ended, and I expect a response from you," "Speak slower" means "Speak slowly," "Say Again" means "Repeat," and "Words twice" means "Give every phrase twice."

Often in radiotelephone communications a "phonetic alphabet" or word list is useful in identifying letters or words that may sound like other letters or words of different meanings. For example "group" may sound like "scoop," or "bridge" may sound like "ridge." A phonetic alphabet or word list consists of a list of 26 words each word beginning with a different letter for identifying that particular letter. If the letters in "Group" are represented in a phonetic word by George, Roger, Oboe, Uncle, and Peter, the "Group" is transmitted as "Group, G as in George, R as in Roger, O as in Oboe, U as in Uncle, P as in Peter."

In making a call by radio, the call sign or name of the called station is generally given 3 times followed by the call letters of the calling station given 3 times.

In testing a radiotelephone transmitter, the operator should clearly indicate that he is testing, and the station call sign or name of the station, as required by the rules, should be clearly given. Tests should be as brief as possible.

If a radio station is used only for occasional calls, it is a good practice to test the station regularly. Regular tests may reveal defects or faults which, if corrected immediately, may prevent delays when communications are necessary. Caution should be observed by persons testing a station to make certain their test message will not interfere with other communications in progress on the same channel. Technical repairs or adjustments to radiotelephone communication stations are made only by or under the immediate supervision and responsibility of operators holding first- and second-class licenses.

When a licensed operator in charge of a radiotelephone station permits another person to use the microphone and talk over the facilities of the station, he should remember that he continues to bear responsibility for the proper operation of the station.

If an operator wishes to determine the specifications for obstruction marking and lighting of antenna towers he should look in Part 17 of the Rules and Regulations of the FCC. If he wishes to determine the specifications for a particular station, he should examine the station authorization issued by the Commission.

1. **What should an operator do when he leaves a transmitter unattended?**—The transmitter should be made inaccessible to unauthorized persons.
2. **What are the meanings of clear, out, over, roger, words twice, repeat, and break?**—Refer to the appropriate words in the summary. "Repeat" means to transmit message or section of message again to make certain it has been copied correctly. "Break" means that the transmitting station will pause briefly for an acknowledgment from the station copying his message, or that the receiving station wishes to break in on the transmitting station to request a repeat.
3. **How should a microphone be treated when used in noisy locations?**—The operator can cup his hand over the microphone to exclude extraneous noise.
4. **What may happen to the received signal when an operator has shouted into a microphone?**—The range of the transmission may be decreased, because the signal may become distorted beyond intelligibility.
5. **Why should radio transmitters be off when signals are not being transmitted?**—Radiation from the transmitter may cause interference even though voice or other information is not being transmitted.
6. **Why should an operator use well-known words and phrases?**—It is advisable to ensure accuracy and save time from undue repetition of words.
7. **Why is the station call sign transmitted?**—So that other stations may clearly identify all calls. Call signs must also be given at prescribed times according to the FCC Rules related to the particular radio service.
8. **Where does an operator find specifications for obstruction marking and lighting where required for the antenna towers of a particular radio station?**—Data may be found in Part 17 of the FCC Rules and Regulations.
9. **What should an operator do if he hears profanity being used at his station?**—He should take steps to prevent the profanity from going out on the air.
10. **When may an operator use a station without regard to certain provisions of the station's license?**—In a period of emergency during which normal communications are disrupted. (R.R. 2.405.)
11. **Who bears the responsibility if an operator permits an unlicensed person to speak over his station?**—The licensed

operator continues to bear responsibility for the proper operation of the station.

12. **What is meant by a phonetic alphabet in radiotelephone communications?**—It is a list of words helpful in identifying letters or words that may sound like other letters or words that have different meanings.

13. **How does the licensed operator of a station normally exhibit his authority to operate the station?**—A valid operator license must be displayed at the transmitter control point.

14. **What precautions should be observed when you are testing a station on the air?**—Be certain testing does not interfere with other communications in progress on the same channel. Clearly indicate that you are testing. *Technical repairs and adjustments* are the responsibility of the holder of a first- or second-class license.

15. **What is the importance of the frequency 2182 kHz?**—This frequency is the international distress frequency for radiotelephony. It can be used by ships, aircraft, and survival craft using frequencies in the 1605- to 4000-kHz spectrum. This frequency is also the international general radiotelephone calling frequency for the maritime mobile service. It may be used by ship stations and aircraft stations operating in the maritime mobile service. (R.R. 83.352, 83.353a.)

16. **Describe completely what action should be taken by a radio operator who hears a distress message; a safety message.**—If he is in the vicinity beyond any possible doubt, the operator should immediately acknowledge receipt. However, in areas where reliable communications with one or more coast stations are practicable, a ship station may defer this acknowledgment for a short interval so that a coast station may acknowledge receipt. If not in the vicinity beyond any possible doubt, the operator shall allow a short interval of time to elapse before acknowledging receipt of the message in order to permit stations nearer to the mobile station in distress to acknowledge receipt of the message without interference.

The acknowledgment of the receipt of a distress message is then made according to established radiotelegraph and radiotelephone operating procedures. On order of the master or person responsible for the ship he is on, the operator shall supply the name of his ship and position and the speed at which it is proceeding toward, and the approximate time it will take to reach the station in distress. However, it is important that the station shall be certain that it will not interfere with the emission of other stations better situated to render immediate assistance to the station in distress.

A station may relay the distress message in any of the following cases: (1) station in distress is not able to transmit a distress message, (2) when a responsible person considers that further help is necessary, (3) when the station is not in position to render assistance but has heard that the distress message has not been acknowledged. Read R.R. 83.240, 83.241, and 83.242 carefully.

A safety signal indicates that the station is about to transmit a message concerning the safety of navigation or giving important meteorological warning. Such a station should be given its proper priority of transmission according to ART. 37. (R.R. 83.239, 83.240, 83.241, 83.242, 83.249, and ART. 37.)

17. **What information must be contained in distress messages? What procedure should be followed by a radio operator in sending a distress message? What is a good choice of words to be used in sending a distress message?** Distress message should contain following information:

    a. Distress call.
    b. Name of ship.
    c. Geographical position.
    d. Nature of distress.
    e. What kind of assistance is needed.
    f. Any additional information that will help in the rescue, such as type of ship, color, length, etc.

The international radiotelegraph distress signal is S O S (. . . - - - . . .) ; the international radiotelephone distress signal is the word "MAYDAY." The radiotelephone distress procedure shall consist of a radiotelephone alarm signal whenever possible, the distress call, and the distress message. This distress transmission shall be made slowly and distinctly, each word being clearly pronounced. Words should be simple and to the point. On advisement, it may be necessary to transmit suitable signals followed by call sign or name to permit direction-finding stations to determine position.

The distress message can be repeated at intervals until an answer is received. If there are no answers, the message may be repeated on any other available frequency on which attention might be attracted.

It is important that you go over each rule and regulation

very carefully, word by word. (R.R. 83.234, 83.235, 83.236, and 83.238.)

18. **What are the requirements for keeping watch on 2182 kHz? If a radio operator is required to stand watch on an international distress frequency, when may he stop listening?**—Each station on board a ship licensed to transmit on telephony within the band 1605 to 3500 kHz shall maintain an efficient watch on 2182 kHz, whenever the station is not being used for transmission on that channel or for communications on other radio channels. Except for stations on board vessels required by law to be fitted with radiotelegraph equipment, each ship station licensed to transmit telephony within the band 1605 to 3500 kHz shall maintain an efficient 2182-kHz watch whenever such station is not being used for transmission on that channel or for communication on other radio channels. (R.R. 83.223.)

19. **Under what circumstances may a coast station contact a land station by radio?**—To facilitate the transmission or reception of safety communications to or from a ship or aircraft station. (R.R. 81.302a2.)

20. **What do distress, urgency, and safety signals indicate? What are the international urgency, safety, and distress signals? In the case of a mobile radio station in distress, what station is responsible for the control of distress-message traffic?**—A distress signal indicates that a mobile station is threatened by grave and eminent danger and requests immediate assistance. The urgency signal indicates that the calling station has a very urgent message to transmit concerning the safety of a ship, aircraft, or other vehicle, or the safety of a person. The safety signal indicates that the station is about to transmit a message concerning the safety of navigation or giving important meteorological warning. These signals are as follows:

	Radiotelegraph	Radiotelephone
Distress	S O S	MAYDAY
Urgency	X X X	PAN
Safety	T T T	SECURITY

Whenever possible the mobile station in distress is responsible for the control of distress-message traffic. (R.R. 83.234 through 83.250.)

21. **In regions of heavy traffic why should an interval be left between radiotelephone calls? Why should a radio operator listen before transmitting on a shared channel? How long may a radio operator in the mobile service continue**

attempting to contact a station which does not answer?
—An interval should be left between radiotelephone calls
to permit the shared use of the channel. An operator should
listen before transmitting on a shared channel to make
certain that he does not interfere with communications in
progress. A call shall not continue for more than 30 seconds. If the called station does not reply, a second call
should not be made until after an interval of two minutes.
If there is no reply to a call sent three times at intervals
of two minutes, the calling shall cease and shall not be
renewed until after an interval of fifteen minutes. However, if there is no reason to believe that harmful interference will be caused, the call sent three times at intervals of
two minutes may be repeated after a pause of not less
than three minutes. (R.R. 83.366 and Summary.)

22. **Why are test transmissions sent? How often should they
be sent? What is the proper way to send a test message?
How often should the station call sign be sent?**—If the
station is used at infrequent intervals, test transmission
can be sent out to ascertain if the transmitter continues to
operate in a normal manner. If the station has not been in
operation, the transmitter can be checked with a test transmission according to an adopted schedule (once each 24
hours is often considered appropriate). When tests are
required, precautions should be taken to prevent any emission that will cause harmful interference. Radiation must
be reduced to the lowest practicable value and entirely suppressed if possible. The licensed operator shall listen carefully before test emission so as not to interfere with transmission in progress.

The official call sign of the testing station followed by the
word "test" shall be announced as a warning that test emissions are to be made. If any other station transmits the
word "wait," testing shall be suspended for an appropriate
interval of time.

During tests the operator then announces the word "testing" followed by a voice transmission test by numerical
count. Test signals should be transmitted that do not conflict with normal operating signals or that will actuate
automatic alarms. The test signal shall have a duration not
exceeding 10 seconds.

At the conclusion of the test a voice announcement of the
official call sign, name of the ship or station, and general
location is to be made. It is customary to give the call sign
three times. The test transmission shall not be repeated

until at least one minute has elapsed; on the frequency 2182 kHz or 156.8 MHz a period of at least five minutes shall elapse before another test transmission is made. (R.R. 83.365 and Summary.)

23. **In the mobile service, why should radiotelephone messages be as brief as possible?**—Transmissions should be brief to give other stations a chance to use the channel and to minimize interference. (Summary.)

24. **What are the meanings of clear, out, over, roger, words twice, repeat, and break?** Refer to the operating summary at the beginning of this chapter.

25. **Does the Geneva 1959 treaty give other countries the authority to inspect US vessels?**—Yes, when the mobile station visits another country. The license or a copy certified by the authority which has issued it should be permanently exhibited in the station. (ART. 21.)

26. **Why are call signs sent? Why should they be sent clearly and distinctly?**—Call signs should be sent and transmitted clearly and distinctly so that the station may be identified exactly by other stations and transmission time kept to a minimum. (Summary.)

27. **How does the licensed operator of a ship station exhibit his authority to operate a station?**—The original license should be exhibited at a conspicuous place at the principal location on board ship. For certain stations of a portable nature and when the operator holds a restricted radiotelephone operator permit, the person may have on his person either the operator license or a verification card. (R.R. 83.165.)

28. **When may a code station not charge for messages it is requested to handle?**—No charge shall be made for the transmission of distress messages and replies thereto in connection with situations involving the safety of life and property at sea, or for any information concerning danger to navigation as designated. Except for effective tariffs on file with the Commission, no charge shall be made for the service of any public coast station (R.R. 81-179.)

29. **What is the difference between calling and working frequencies?**—Calling frequencies are used for transmissions from a station solely to secure the attention of another station for a particular purpose. A working frequency is a frequency that is used strictly for carrying on radiocommunications for a purpose other than calling by any station or stations using telegraphy, telephony, or facsimile. (R.R. 83.6.)

ELEMENTS I AND II SELF-TEST

1. **When a licensee qualifies for a higher grade license, the lesser grade license**
   - A. is cancelled.
   - B. must be returned to the FCC.
   - C. is still valid.
   - D. may be used as a verification card.

2. **When a license or permit is lost the licensee**
   - A. must notify FCC within 30 days.
   - B. must submit application for a duplicate immediately.
   - C. may not operate.
   - D. is dropped to a lower grade.

3. **Radio station inspection is done by the**
   - A. FBI.
   - B. FCC.
   - C. FAA.
   - D. CIA.

4. **Station logs are signed**
   - A. by FCC inspectors.
   - B. when new equipment is installed.
   - C. at local sunrise.
   - D. when starting and going off duty.

5. **An error in the log is corrected by**
   - A. erasure and dating.
   - B. erasure and signing by the person correcting the log.
   - C. rewriting the entire page.
   - D. striking, dating, and signing by the person correcting the log.

6. **Profanity on the air**
   - A. is legal.
   - B. is a violation of FCC rules and regulations.
   - C. is permitted because it emphasizes a situation.
   - D. must be used on the proper frequency.

7. **The urgency signal has**
   - A. more priority than a distress signal.
   - B. more priority than a security signal.
   - C. less priority than a safety signal.
   - D. a TTT radiotelegraph signal.

8. **An operator may divulge the contents of a**
   - A. distress message.
   - B. business transaction.
   - C. marine radiotelephone conversation.
   - D. commercial shipping instructions.

9. **When a licensee receives a notice of suspension it shall take effect**
   - A. immediately.
   - B. in 15 days.
   - C. in 30 days.
   - D. in 24 hours.

10. A spurious emission that interrupts a radiocommunication service

A. is permissible at low power.
B. is permissible on shared channels.

C. is a strong cw signal.
D. is harmful interference.

11. "Over" means

A. change over to a new frequency.
B. change to another mode of transmission.

C. change cw speed.
D. my transmission is ended and a reply is expected.

12. Overmodulation can be caused by

A. shouting into the microphone.
B. operating on a wrong frequency.

C. too little modulating power.
D. posting a wrong license.

13. Accuracy is improved by

A. using long descriptive words and sentences.
B. talking fast.

C. using well-known words.
D. shouting into the microphone.

14. Lights of an antenna tower should be inspected every

A. 24 hours.
B. 48 hours.

C. week.
D. 30 days.

15. A coast station may contact a land station

A. for a friendly chat.
B. to handle ship traffic.

C. to aid in transmission of safety messages.
D. to relay press traffic.

16. During an emergency

A. every station should get on the air.
B. certain provisions of a station license need not be followed when normal communications are disrupted.

C. shout into the microphone.
D. chase distress station off your channel.

17. If in operating on a busy channel you cannot contact the desired station after 7½ minutes of calling,

A. wait 15 minutes or until propagation conditions improve.
B. increase power above legal limit.

C. shout louder into microphone.
D. change over to sideband.

18. **The phonetic language**

    A. is seldom used in radio-communications.
    B. helps to clarify phrases.
    C. is a method of clarifying cw.
    D. helps to clarify letters.

19. **When testing**

    A. do it in a hurry and don't waste time listening.
    B. always use maximum assigned power.
    C. be certain testing does not interfere with communications in progress.
    D. shout into the microphone.

20. **When you receive a distress message**

    A. throw your carrier on the air and be ready to help.
    B. try to keep everyone off your channel.
    C. be certain you will not interfere with stations better situated to render assistance.
    D. shut down your station.

21. **An operator required to stand watch on an international distress frequency may stop listening**

    A. when he is handling traffic on another channel.
    B. when it is dark.
    C. when he is 300 miles from shore.
    D. on legal holidays.

22. **The radiotelephone urgency signal is**

    A. MAYDAY.
    B. PAN.
    C. SECURITY.
    D. XXX.

23. **The proper way to start a test message on a clear channel is**

    A. turn on transmitter quickly and shout TEST.
    B. repeat word TESTING and call sign for 20 seconds.
    C. give station call sign and word TEST; then listen on channel before proceeding.
    D. call a friend and tell him to keep channel open.

24. **The word "clear" when used at end of transmission means**

    A. there is to be clear weather.
    B. the channel is free of signals.
    C. propagation is good.
    D. communications have been concluded with contacted station.

25. **Call signs**

    A. are a fad of radio operators.
    B. help to identify stations exactly.
    C. can be changed by operator when he gets tired of old one.
    D. are a useless tradition.

26. **Who may make application for a radio operator's license?**

    A. Citizens.
    B. Nationals of the United States.
    C. FAA-certified alien pilots.
    D. All of the above.

27. **An operator's license may be renewed**

A. a year before expiration.
B. within a year after expiration.
C. within two years after expiration.
D. as in A and B above.

28. **If your channel is active and you have business traffic**

A. put your carrier on the air.
B. tell others to get off the channel.
C. you may not deliberately interfere with communications.
D. call the FCC and say you want a new channel.

29. **Which is not ground for license suspension?**

A. Damaging radio apparatus.
B. Use of indecent language.
C. Failure to carry out orders of master of ship.
D. Overmodulation.

30. **If a licensee is informed he has violated an FCC rule he must reply within**

A. 24 hours.
B. a week.
C. 10 days.
D. 30 days.

31. **What are the maximum penalties for violating the Communications Act?**

A. $500 or one year imprisonment.
B. $10,000 or two-years imprisonment or both.
C. $5000 or one-year imprisonment or both.
D. $500.

32. **"Words Twice" means**

A. repeat every word one by one.
B. give every phrase twice.
C. use phonetic language.
D. repeat every letter twice.

33. **When operating in a noisy location**

A. shout louder into the microphone.
B. cup the hands around the microphone.
C. increase power.
D. change channel.

34. **Where does an operator find specifications for obstruction lights on a radio tower?**

A. Part 17 FCC Rules and Regulations.
B. FAA office.
C. Local authority.
D. Local airport.

35. **A call to another station should not continue for more than**

A. 30 seconds.
B. 20 seconds.
C. 1 minute.
D. 2 minutes.

36. An assigned call or identifier must be transmitted

A. at the end of each
transmission or exchange.
B. every 30 minutes.

C. only when convenient.
D. either A or B.

37. You are called upon to service many stations in the Land-Mobile Services. How can you do the work legally with only one posted license?

A. Use any number of
photostatic copies.
B. Use verification card 758F.

C. Use driver's license for
identification.
D. Carry station license.

38. Who may not sign an application for a station license?

A. Individual making
application.
B. Responsible officer for a
station to be operated by
a company.

C. Responsible appointed official for a local government
station.
D. A friend of applicant who
has a first-class license.

39. In periods of emergency

A. certain provisions of station
license need not be regarded.
B. you may not transmit under
any circumstances

C. you must call FCC district
office.
D. reduce transmitter power.

40. All license applications must be

A. signed by the applicant.
B. notarized.

C. sent to National Bureau
of Standards.
D. sent to FAA

ANSWERS TO SELF-TEST

1. A.
2. B. If the original is found meanwhile, it is necessary that you return either one, original or duplicate, to the FCC.
3. B.
4. D. Station logs must be signed when the operator goes on and off duty. Likewise log records must be signed by the licensed maintenance technician and must also include data on the nature of the transmitter repairs.
5. D.
6. B.
7. B.
8. A.
9. B.
10. D.
11. D.
12. A.
13. C.
14. A.
15. C.
16. B.
17. A. The operator must learn to be patient under adverse propagation conditions. Repeated calls should be made only if one is certain the transmissions cause no interference or hamper other communications that share the same frequency.
18. D.
19. C.
20. C.
21. A.

22. B.
23. C.
24. D.
25. B.
26. D.
27. D. Questions 26 and 27 show a
method of questioning that is
often used in FCC examinations. You are to select the most
complete answer. In Question
27, both A and B are correct
but the proper answer is D in
terms of multiple-choice evalu-
ation.
28. C.
29. D. Overmodulation is to be
avoided because it limits the
effective transmission range of
the station due to speech dis-
tortion. However, it is not direct grounds for license suspension.
30. C.
31. B.
32. B.
33. B. This is a problem when one
must transmit from a location
near operating machinery or in
traffic. The development of
noise-cancelling microphones
has helped to overcome this
problem. They must be close-
talked to be really useful. Close-
talk, but do not shout.
34. A.
35. A.
36. D.
37. B.
38. D.
39. A.
40. A.

APPENDIX A

Extracts From the Geneva, 1959, Treaty

SECTION III

TECHNICAL CHARACTERISTICS

PARAGRAPH 93. Harmful interference: Any emission, radiation or induction which endangers the functioning of radionavigation service or of other safety services or seriously degrades, obstructs or repeatedly interrupts a radiocommunication service operating in accordance with these Regulations.

ARTICLE 21

INSPECTION OF MOBILE STATIONS

838 SEC. 1. (1) The governments or appropriate administrations of countries which a mobile station visits, may require the production of the license for examination. The operator of the mobile station, or the person responsible for the station, shall facilitate this examination. The license shall be kept in such a way that it can be produced upon request. As far as possible, the license, or a copy certified by the authority which has issued it, should be permanently exhibited in the station.

ARTICLE 37

PARAGRAPH 1496. The term "communication" as used in this Article means radiotelegrams as well as radiotelephone calls. The order of priority for communications in the mobile service shall be as follows:

1. Distress calls, distress messages, and distress traffic.
2. Communications preceded by the urgency signal.
3. Communications preceded by the safety signal.
4. Communications relating to radio direction-finding.
5. Communications relating to the navigation and safe movement of aircraft.
6. Communications relating to the navigation, movements, and needs of ships, and weather observation messages destined for an official meteorological service.
7. Government radiotelegrams: Priorite' Nations.
8. Government communications for which priority has been requested.
9. Service communications relating to the working of the radio-communications previously exchanged.
10. Government communications other than those shown in 7 and 8 above, and all other communications.

Extracts From the Communications Act of 1934, as Amended

SEC. 303 Except as otherwise provided in this Act, the Commission from time to time, as public convenience, interest, or necessity requires shall—(m) (1) Have authority to suspend the license of any operator upon proof sufficient to satisfy the Commission that the licensee

(A) Has violated any provision of any Act, treaty, or convention binding on the United States, which the Commission is authorized to administer, or any regulation made by the Commission under any such Act, treaty, or convention; or

(B) Has failed to carry out a lawful order of the master or person lawfully in charge of the ship or aircraft on which he is employed; or

(C) Has willfully damaged or permitted radio apparatus or installations to be damaged; or

(D) Has transmitted superfluous radio communications or signals or communications containing profane or obscene words, language, or meaning, or has knowingly transmitted

    (1) False or deceptive signals or communications, or

    (2) A call signal or letter which has not been assigned by proper authority to the station he is operating; or

(E) Has willfully or maliciously interfered with any other radio communications or signals; or

(F) Has obtained or attempted to obtain, or has assisted another to obtain or attempt to obtain, an operator's license by fraudulent means.

*   *   *   *   *   *   *

(n) Have authority to inspect all radio installations associated with stations required to be licensed by any Act, or which are subject to the provisions of any Act, treaty, or convention binding on the United States, to ascertain whether in construction, installations, and operation they conform to the requirements of the rules and regulations of the Commission, the provisions of any Act, the terms of any treaty or convention binding on the United States, and the conditions.

*   *   *   *   *   *   *

SEC. 325 (a)   No person within the jurisdiction of the United States shall knowingly utter or transmit, or cause to be uttered or transmitted, any false or fraudulent signal of distress, or communication relating thereto, nor shall any broadcasting station rebroadcast the program or any part thereof of another broadcasting station without the express authority of the originating station.

*   *   *   *   *   *   *

SEC. 501   Any person who willfully and knowingly does or causes or suffers to be done any act, matter, or thing, in this Act prohibited or declared to be unlawful, or who willfully or knowingly omits or fails to do any act, matter, or thing in this Act required to be done, or willfully and knowingly causes or suffers such omission or failure, shall, upon conviction thereof, be punished for such offense, for which no penalty (other than a forfeiture) is provided in this Act, by a fine of not more than $10,000 or by imprisonment for a term not exceeding one year, or both; except that any person, having been once convicted of an offense punishable under this section, who is subsequently convicted of violating any provision of this Act punishable under this section, shall be punished by a fine of not more than $10,000 or by imprisonment for a term not exceeding two years; or both.

SEC. 502   Any person who willfully and knowingly violates any rule, regulation, restriction, or condition made or imposed by the Commission under authority of this Act, or any rule, regulation, restriction, or condition made or imposed by any international radio or wire communications treaty or convention, or regulations annexed thereto, to which the United States is or may hereafter become a party, shall, in addition to any other penalties provided by law, be punished, upon conviction thereof, by a fine of not more than $500 for each and every day during which such offense occurs.

*   *   *   *   *   *   *

SEC. 605   No person receiving or transmitting, or assisting in transmitting, any interstate or foreign communication by wire or radio shall divulge or publish the existence, contents, substance, purport, effect, or meaning thereof, except through authorized channels of transmission or reception, to any person other than the addressee, his agent, or attorney, or to a person employed or authorized to forward such communication to its destination, or to proper accounting or distributing officers of the various communicating centers over which the communication may be passed, or to the master of a ship under whom he is serving, or in response to a subpoena issued by a court of competent jurisdiction, or on demand of other lawful authority; and no person not being authorized by the sender shall intercept any communication and divulge or publish the existence, contents, substance, purport, effect, or meaning of such intercepted communication to any person; and no person not being entitled thereto shall receive or assist in receiving any interstate or foreign communication by wire or radio and use the same or any information therein contained for his own benefit or for the benefit of another not entitled thereto; and no person having received such intercepted communication or having become acquainted with the contents, substance, purport, effect, or meaning of the same or any part thereof, knowing that such information was so obtained, shall divulge or publish the existence, contents, substance, purport, effect, or meaning of the same or any part thereof, or use the same or any information therein contained for his own benefit or for the benefit of another not entitled thereto: *Provided*, That this section shall not apply to the receiving, divulging, publishing, or utilizing the contents of any radio communication broadcast, or transmitted by amateurs or others for the use of the general public, or relating to ships in distress.

APPENDIX C

Extracts From the FCC
Rules and Regulations,
Parts 1, 2, 13, 17, 81, and 83

This appendix contains reproductions of portions of Parts 1, 2, 13, 17, 81, and 83 of the FCC Rules and Regulations. Even though the information was current at the time this book was printed, the radio operator should have access to and be familiar with a current copy of the Rules and Regulations, since they are subject to revision.

§ 1.85 Suspension of operator licenses.

Whenever grounds exist for suspension of an operator license, as provided in section 303(m) of the Communications Act, the Chief of the Safety and Special Radio Services Bureau, with respect to amateur operator licenses, or the Chief of the Field Engineering Bureau, with respect to commercial operator licenses, may issue an order suspending the operator license. No order of suspension of any operator's license shall take effect until 15 days' notice in writing of the cause for the proposed suspension has been given to the operator licensee, who may make written application to the Commission at any time within said 15 days for a hearing upon such order. The notice to the operator licensee shall not be effective until actually received by him, and from that time he shall have 15 days in which to mail the said application. In the event that physical conditions prevent mailing of the application before the expiration of the 15-day period, the application shall then be mailed as soon as possible thereafter, accompanied by a satisfactory explanation of the delay. Upon receipt by the Commission of such application for hearing, said order of suspension shall be designated for hearing by the Chief, Safety and Special Radio Services Bureau, or the Chief, Field Engineering Bureau, as the case may be, and said order of suspension shall be held in abeyance until the conclusion of the hearing. Upon the conclusion of said hearing, the Commission may affirm, modify, or revoke said order of suspension. If the license is ordered suspended, the operator shall send his operator license to the office of the Commission in Washington, D.C., on or before the effective date of the order, or, if the effective date has passed at the time notice is received, the license shall be sent to the Commission forthwith.

(Sec. 303(m), 48 Stat. 1082, as amended; 47 U.S.C. 303(m))

* * * * * * *

§ 1.89 Notice of violations.

(a) Except in cases of willfulness or those in which public health, interest, or safety requires otherwise, any licensee who appears to have violated any provision of the Communications Act or any provision of this chapter will, before revocation, suspension, or cease and desist proceedings are instituted, be served with a written notice calling these facts to his attention and requesting a statement concerning the matter. FCC Form 793 may be used for this purpose. The Notice of Violation may be combined with a Notice of Apparent Liability to Monetary Forfeiture. In such event, notwithstanding the Notice of Violation, the provisions of § 1.80 apply and not those of § 1.89.

(b) Within 10 days from receipt of notice or such other period as may be specified, the licensee shall send a written answer, in duplicate, direct to the office of the Commission originating the official notice. If an answer cannot be sent nor an acknowledgment made within such 10-day period by reason of illness or other unavoidable circumstances, acknowledgment and answer shall be made at the earliest practicable date with a satisfactory explanation of the delay.

(c) The answer to each notice shall be complete in itself and shall not be abbreviated by reference to other communications or answers to other notices. In every instance the answer shall contain a statement of action taken to correct the condition or omission complained of and to preclude its recurrence. In addition:

(1) If the notice relates to violations that may be due to the physical or electrical characteristics of transmitting apparatus and any new apparatus is to be installed, the answer shall state the date such apparatus was ordered, the name of the manufacturer, and the promised date of delivery. If the installation of such apparatus requires a construction permit, the file number of the application shall be given, or if a file number has not been assigned by the Commission, such identification shall be given as will permit ready identification of the application.

(2) If the notice of violation relates to lack of attention to or improper operation of the transmitter, the name and license number of the operator in charge shall be given.

*   *   *   *   *   *   *

§ 2.405 Operation during emergency.

The licensee of any station (except amateur, standard broadcast, FM broadcast, noncommercial educational FM broadcast, or television broadcast) may, during a period of emergency in which normal communication facilities are disrupted as a result of hurricane, flood, earthquake, or similar disaster, utilize such station for emergency communication service in communicating in a manner other than that specified in the instrument of authorization: *Provided:* (a) That as soon as possible after the beginning of such emergency use, notice be sent to the Commission at Washington, D.C., and to the Engineer in Charge of the district in which the station is located, stating the nature of the emergency and the use to which the station is being put, and (b) That the emergency use of the station shall be discontinued as soon as substantially normal communication facilities are again available, and (c) That the Commission at Washington, D.C., and the Engineer in Charge shall be notified immediately when such special use of the station is terminated: *Provided further,* (d) That in no event shall any station engage in emergency transmission on frequencies other than, or with power in excess of, that specified in the instrument of authorization or as otherwise expressly provided by the Commission, or by law: *And provided further,* (e) That any such emergency communication undertaken under this section shall terminate upon order of the Commission.

*   *   *   *   *   *   *

§ 13.4 Term of licenses.

(a) Except as provided in paragraphs (b) and (c) of this section, commercial operator licenses will normally be issued for a term of 5 years from the date of issuance.

(b) Restricted Radiotelephone Operator Permits issued to U.S. citizens or other U.S. nationals will normally be issued for the lifetime of the operator. The terms of all Restricted Radiotelephone Operator Permits issued prior to November 15, 1953, which were outstanding on that date were extended to encompass the lifetime of such operators.

(c) A commercial operator license or permit granted to an alien aircraft pilot under a waiver of the United States nationality provisions of Section 303(1) of the Communications Act will normally be issued for a term of five (5) years from the date of issuance. An operator license or permit issued to an alien aircraft pilot shall be valid only if the operator continues to hold a valid aircraft pilot certificate issued by the Federal Aviation Administration or one of its predecessor agencies.

§ 13.5 Eligibility for new license.

(a) Commercial operator licenses are issued to

(1) United States citizens.

(2) United States nationals.

(3) Citizens of the Trust Territory of the Pacific Islands presenting valid identity certificates issued by the High Commissioner of the Trust Territory.

(4) Aliens holding Federal Aviation Administration pilot certificates.

(5) Aliens holding Federal Communications Commission station licenses, but the operator license will be valid only for operating the station licensed in the name of the alien.

*   *   *   *   *   *   *

§ 13.6 Operator license, posting of.

The original license of each station operator shall be posted at the place where he is on duty, except as otherwise provided in this part or in the rules governing the class of station concerned.

*   *   *   *   *   *   *

§ 13.7 Operators, place of duty.

(a) Except as may be provided in the rules governing a particular class of station, one or more licensed radio operators of the grade specified by this part shall be on duty at the place where the transmitting apparatus of each licensed radio station is located and in actual charge thereof whenever it is being operated: *Provided, however,* That (1) subject to the provisions of paragraph (b) of this section, where remote control of the transmitting apparatus has been authorized to be used, the Commission may modify the foregoing requirements upon proper application and showing being made so that such operator or operators may be on duty at the control point in lieu of the place where the transmitting apparatus is located; (2) in the case of two or more stations, except broadcast, licensed in the name of the same person to use frequencies above 30 megahertz only, a licensed radio operator holding a valid radiotelegraph or radiotelephone first- or second-class license who has the station within his effective control may be on duty at any point within the communication range of such stations in lieu of the transmitter location or control point during the actual operation of the transmitting apparatus and shall supervise the emissions of all such stations so as to insure the proper operation in accordance with the station license.

(b) An operator may be on duty at a remote control point in lieu of the location of the transmitting apparatus in accordance with the provisions of paragraph (a) (1) of this section: *Provided,* That all of the following conditions are met: (1) The transmitter shall be so installed and protected that it is not accessible to other than duly authorized persons; (2) the emissions of the transmitter shall be continuously monitored at the control point by a licensed operator of the grade specified for the class of station involved; (3) provision shall be made so that the transmitter can quickly and without delay be placed in an inoperative condition by the operator at the control point in the event there is a deviation from the terms of the station license; (4) the radiation of the transmitter shall be suspended immediately when there is a deviation from the terms of the station license.

§ 13.8 Provisional Radio Operator Certificate.

(a) In circumstances requiring immediate authority to operate a radio station pending submission of proof of eligibility or of qualifications or pending a determination by the Commission as to these matters, an applicant for a radio operator license may be issued a Provisional Radio Operator Certificate.

(b) If the Commission finds that the public interest will be served, it may issue such a certificate for a period not to exceed 6 months with such additional limitations as may be indicated.

(c) A Provisional Radio Operator Certificate will not be issued if the applicant has not fulfilled the examination requirements for the license applied for.

(d) [Reserved]

APPLICATIONS

§ 13.11 Procedure.

(a) *General.* Applications shall be governed by applicable rules in force on the date when application is filed (see § 13.28). The application in the prescribed form and including all required subsidiary forms and documents, properly completed and signed, and accompanied by the fee prescribed in the Commission's general fee schedule as set forth in Subpart G, Part 1 of this chapter, shall be submitted to the appropriate office as indicated in paragraph (b) of this section. If the application is for renewal of license, it may be filed at any time during the final year of the license term or during a 1-year period of grace after the date of expiration of the license sought to be renewed. During this 1-year period of grace, an expired license is not valid. A renewed license issued upon the basis of an application filed during the grace period will be dated currently and will not be backdated to the date of expiration of the license being renewed. A renewal application shall be accompanied by the license sought to be renewed. If the prescribed service requirements for renewal without examination (see § 13.28) are fulfilled, the renewed license may be issued by mail. If the service record on the reverse side of the license does not fully describe or cover the service desired by the applicant to be considered in connection with license renewal (as might occur in the case of service rendered at U.S. Government stations), the renewal application shall be supported by documentary evidence describing in detail the service performed and showing that the applicant actually performed such service in a satisfactory manner. A separate application must be submitted for each license involved, whether it requests renewal, new license, endorsement, duplicate, or replacement.

(b) *Place of filing.* (1) Applications for Restricted Radiotelephone Operator Permits shall be filed as follows:

(i) United States citizens, United States nationals, citizens of the Trust Territory of the Pacific Islands—File application FCC Form 753:

(a) When there is an immediate need for the permit for safety purposes, submit the application in person or by an agent to the nearest field office of the Field Operations Bureau—Federal Communications Commission. The application must be accompanied by a written showing by the applicant that he has an immediate need for the permit for safety purposes.

(b) When the applicant is located in Alaska, Hawaii, Puerto Rico, or the Virgin Islands of the United States, submit the application in person or by mail to the nearest field office of the Field Operations Bureau—Federal Communications Commission.

(c) All other cases—mail the application to the Federal Communications Commission, Gettysburg, Pennsylvania 17325.

(ii) Aliens holding a station license—File application FCC Form 755:

(a) When there is an immediate need for the permit for safety purposes, submit the application in person or by an agent to the nearest field office of the Field Operations Bureau—Federal Communications Commission. The application must be accompanied by a written showing by the applicant that he has an immediate need for the permit for safety purposes.

(b) When the applicant is located in Alaska, Hawaii, Puerto Rico, or the Virgin Islands of the United States, submit the application in person or by mail to the nearest field office of the Field Operations Bureau—Federal Communications Commission.

(c) All other cases—mail application to the Federal Communications Commission, Washington, D.C. 20554.

(iii) Alien aircraft pilots holding Federal Aviation Administration pilot certificates—Submit the application FCC Form 755 ether in person or by mail to the Federal Communications Commission, Washington, D.C. 20554 (see § 13.4(c)).

(2) An application for an operator license or permit of any other class, or for a verification card, shall be submitted in person or by mail to the field office at which the applicant desires his application to be considered and acted upon, and which office will make final arrangements for conducting any required examination. Whenever an examination is required to be taken at a designated examination point away from a field office, the application shall be submitted in advance of the examination to the field office having jurisdiction over the area in which the examination is to be given.

(3) The form entitled "Verification of Operator License or Permit" (FCC 759) may be obtained from any of the Commission's field offices. The certification under Part B of the form shall be completed by the licensee or general manager of the radio station where the statement is to be posted. When the FCC Form 759 is properly validated, it may be posted in lieu of the original radio operator license or permit when the holder of that license or permit is employed at more than one station.

(c) *Restricted radiotelephone operator permit.* No oral or written examination is required for this permit. If the application is properly completed and signed, and if the applicant is found to be qualified, the permit may be issued forthwith.

(d) *Short term license.* A license or permit issued for a term of less than five years (see § 13.4), may be renewed without further examination, provided proper application is filed in accordance with paragraph (a) of this section.

(e) *Blind applicant.* A blind person seeking an examination for radiotelephone first class operator license, radiotelephone second class operator license, or radiotelephone third class operator license shall make a request in writing to the appropriate field office for a time and date to appear for such examination. The examination shall be administered only at the field office. Requests for examinations shall be made at least 2 weeks prior to the date on which the examination is desired.

[§ 13.11(e) amended eff. 2-7-79; I(77)-5]

§ 13.12 Special provisions, radiotelegraph first class.

An applicant for a radiotelegraph first-class operator license must be at least 21 years of age at the time the license is issued and shall have had an aggregate of one year of satisfactory service as an operator manipulating the key of a manually operated public ship or coast station handling public correspondence by radiotelegraphy.

§ 13.13 Age limit, restricted radiotelephone operator permit.

An applicant for a restricted radiotelephone operator permit must be at least 14 years of age at the time the permit is issued.

EXAMINATIONS

§ 13.21 Examination elements.

(a) Written examinations will comprise questions from one or more of the following examination elements.

(1) Basic law. Provisions of laws, treaties and regulations with which every operator should be familiar.

(2) Basic operating practice. Radio operating procedures and practices generally followed or required in communicating by means of radiotelephone stations.

(3) Basic radiotelephone. Technical, legal and other matters applicable to the operation of radiotelephone stations other than broadcast.

(4) Advanced radiotelephone. Advanced technical, legal and other matters particularly applicable to the operation of the various classes of broadcast stations.

(5) Radiotelegraph operating practice. Radio operating procedures and practices generally followed or required in communicating by means of radiotelegraph stations primarily other than in the maritime mobile services of public correspondence.

(6) Advanced radiotelegraph. Technical, legal and other matters applicable to the operation of all classes of radiotelegraph stations, including operating procedures and practices in the maritime mobile services of public correspondence, and associated matters such as radio navigational aids, message traffic routing and accounting, etc.

(7) Aircraft radiotelegraph. Basic theory and practice in the operation of radio communications and radio navigational systems in general use on aircraft.

(8) Ship radar techniques. Specialized theory and practice applicable to the proper installation, servicing and maintenance of ship radar equipment in general use for marine navigational purposes.

(b) Examination elements of the radiotelegraph Morse Code comprise sending and receiving tests at the following rates:

1. Sixteen (16) code groups-per-minute.
2. Twenty (20) plain language words-per-minute.
3. Twenty (20) code groups-per-minute.
4. Twenty-five (25) plain language words-per-minute.

*   *   *   *   *   *   *

§ 13.22 Examination requirements.

Applicants for original licenses will be required to pass examinations as follows:

(a) *Radiotelephone second-class operator license:*

(1) Ability to transmit and receive spoken messages in English.

(2) Written examination elements: 1, 2, and 3.

(b) *Radiotelephone first-class operator license:*

(1) Ability to transmit and receive spoken messages in English.

(2) Written examination elements: 1, 2, 3, and 4.

(c) *Radiotelegraph second-class operator license:*

(1) Ability to transmit and receive spoken messages in English.

(2) Transmitting and receiving code test of twenty (20) words per minute plain language and sixteen (16) code groups per minute.

(3) Written examination elements: 1, 2, 5, and 6.

(d) [Reserved]

(e) *Radiotelegraph first-class operator license:*

(1) Ability to transmit and receive spoken messages in English.

(2) Transmitting and receiving code test of twenty-five (25) words per minute plain language and twenty (20) code groups per minute.

(3) Written examination elements: 1, 2, 5, and 6.

(f) *Radiotelephone third-class operator permit:*

(1) Ability to transmit and receive spoken messages in English.

(2) Written examination elements: 1 and 2.

(g) *Radiotelegraph third-class operator permit:*

(1) Ability to transmit and receive spoken messages in English.

(2) Transmitting and receiving code test of twenty (20) words per minute plain language and sixteen (16) code groups per minute.

(3) Written examination elements: 1, 2, and 5.

(h) *Restricted radiotelephone operator permit:*

No oral or written examination is required for this permit. In lieu thereof, applicants will be required to certify in writing to a declaration which states that the applicant has need for the requested permit; can receive and transmit spoken messages in English; can keep at least a rough written log in English or in some other language in general use that can be readily translated into English; is familiar with the provisions of treaties, laws, and rules and regulations governing the authority granted under the requested permit; and understands that it is his responsibility to keep currently familiar with all such provisions.

§ 13.23 Examination form.

The written examination shall be in English, except when waived in accordance with authority specified in section 0.314. In the case of a blind applicant, the examination questions shall be read orally and the dictated answers recorded by a Commission examiner authorized to administer such oral examination.

§ 13.24 Passing mark.

A passing mark of 75 percent of a possible 100 percent will be required on each element of a written examination.

§ 13.25 New class, additional requirements.

The holder of a license, who applies for another class of license, will be required to pass only the added examination requirements for the new class of license: *Provided,* That the holder of a radiotelegraph third-class operator permit who takes an examination for a radiotelegraph second-class operator license more than one year after the issuance date of the third-class permit will also be required to pass the code test prescribed therefor: *Provided further.* That no person holding a new, duplicate, or replacement restricted radiotelephone operator permit issued on the basis of a declaration, or a renewed restricted radiotelephone operator permit which renews a permit issued upon the basis of a declaration shall, by reason of the declaration or the holding of such permit, be relieved in any respect of qualifying by examination when applying for any other class of license.

§ 1326 Canceling and issuing new licenses.

If the holder of a license qualifies for a higher class in the same group, the license held will be canceled upon the issuance of the new license. Similarly, if the holder of a restricted operator permit qualifies for any other class or type of license or permit, the Restricted Radiotelephone Operator Permit will be canceled upon issuance of the new authorization, except as provided in section 13.3.

§ 13.27 Eligibility for re-examination.

An applicant who fails an examination element, including a code test element, will be ineligible for 2 months to take an examination for any class of license requiring that element. Examination elements will be graded in the order listed (see § 13.21), and an applicant may, without further application, be issued the class of license for which he qualifies.

NOTE: A month after date is the same day of the following month, or if there is no such day, the last day of such month. This principle applies for other periods. For example, in the case of the 2-month period to which this note refers, an applicant examined December 1 may be reexamined February 1, and an applicant examined December 29, 30, or 31 may be reexamined the last day of February while one examined February 28 may be reexamined April 28.

§ 13.28 Renewal service requirements, renewal examinations, and exceptions.

A restricted radiotelephone operator permit normally is issued for the lifetime of the holder and need not be renewed, EXCEPT that alien restricted radiotelephone operator permits are normally issued for a five year term and are normally renewable. A license of any other class may be renewed without examination provided that the service record on the reverse side of the license (see §§ 13.91 to 13.94) shows at least two years of satisfactory service in the aggregate during the license term and while actually employed as a radio operator under the license. If this two-year renewal service requirement is not fulfilled, but the service record shows at least one year of satisfactory service in the aggregate during the last three years of the license term and while actually employed as a radio operator under that license, the license may be renewed upon the successful completion of a renewal examination, which may be taken at any time during the final year of the license term or during a one-year period of grace after the date of expiration of the license sought to be renewed. The renewal examination will consist of the highest num-bered examination element normally required for a new license of the class sought to be renewed, plus the code test (if any) required for such a new license. If the renewal examination is not successfully completed before expiration of the aforementioned one-year period of grace, the license will not be renewed on any basis.

NOTE: By order dated and effective April 4, 1951, the Commission temporarily waived the requirement of prior service as a radio operator or examination for renewal in the case of any applicant for renewal of his commercial radio operator license. This order is applicable to commercial radio operator licenses which expired after June 30, 1950 until further order of the Commission.

*   *   *   *   *   *   *

CODE TESTS

§ 13.41 Transmitting speed requirements.

An applicant is required to transmit correctly in the International Morse code for 1 minute at the rate of speed prescribed in this part for the class of license desired.

§ 13.42 Transmitting test procedure.

Transmitting tests shall be performed by the use of the conventional Morse key except that a semiautomatic key, if furnished by the applicant, may be used in transmitting code tests of 25 words per minute.

§ 13.43 Receiving speed requirements.

An applicant is required to receive the International Morse code by ear, and legibly transcribe, consecutive words or code groups for a period of 1 minute without error at the rate of speed specified in the rules for the class of license for which the application is made.

§ 13.44 Receiving test procedure.

Receiving code tests shall be written in longhand or printed by hand either in ink or pencil except that in the case of the 25 words per minute code test a typewriter may be used when furnished by the applicant.

§ 13.45 Computing words or code groups.

Each five characters shall be counted as one word or code group. Punctuation marks or figures count as two characters.

SCOPE OF AUTHORITY

§ 13.61 Operating authority.

The various classes of commercial radio operator licenses issued by the Commission authorize the holders thereof to operate radio stations, except amateur, as follows (See also § 13.62(c) for additional operating authority with respect to standard and FM broadcast stations):

(a) *Radiotelegraph first-class operator license.* Any station except:

(1) Stations transmitting television, or

(2) Any of the various classes of broadcast stations, or

(3) On a cargo vessel (other than a vessel operated exclusively on the Great Lakes) required by treaty or statute to be equipped with a radiotelegraph installation, the holder of this class of license may not act as chief or sole operator until he has had at least 6 months' satisfactory service in the aggregate as a qualified radiotelegraph operator in a station on

board a ship or ships of the United States.

(4) On an aircraft employing radiotelegraphy, the holder of this class of license may not operate the radiotelegraph station during the course of normal rendition of service unless he has satisfactorily completed a supplementary examination qualifying him for that duty, or unless he has served satisfactorily as chief or sole radio operator on an aircraft employing radiotelegraphy prior to February 15, 1950. The supplementary examination shall consist of:

(1) Written examination element: 7.

(5) At a ship radar station, the holder of this class of license may not supervise or be responsible for the performance of any adjustments or tests during or coincident with the installation, servicing or maintenance of the radar equipment while it is radiating energy unless he has satisfactorily completed a supplementary examination qualifying him for that duty and received a ship radar endorsement on his license certifying to that fact: *Provided,* That nothing in this subparagraph shall be construed to prevent persons holding licenses not so endorsed from making replacements of fuses or of receiving-type tubes. The supplementary examination shall consist of:

(i) Written examination element: 8.

(b) *Radiotelegraph second-class operator license.* Any station except:

(1) Stations transmitting television, or

(2) Any of the various classes of broadcast stations, or

(3) On a passenger vessel (a ship shall be considered a passenger ship if it carries or is licensed or certificated to carry more than 12 passengers; a cargo ship means any ship not a passenger ship) required by treaty or statute to maintain a continuous radio watch by operators or on a vessel having continuous hours of service for public correspondence, the holder of this class of license may not act as chief operator, or

(4) On a vessel (other than a vessel operated exclusively on the Great Lakes) required by treaty or statute to be equipped with a radiotelegraph installation, the holder of this class of license may not act as chief or sole operator until he has had at least 6 months' satisfactory service in the aggregate as a qualified radiotelegraph operator in a station on board a ship or ships of the United States.

(5) On an aircraft employing radiotelegraphy, the holder of this class of license may not operate the radiotelegraph station during the course of normal rendition of service unless he is at least eighteen (18) years of age and has satisfactorily completed a supplementary examination qualifying him for that duty, or unless he has served satisfactorily as chief or sole radio operator on an aircraft employing radiotelegraphy prior to February 15, 1950. The supplementary examination shall consist of:

(i) Transmitting and receiving code test at twenty-five (25) words per minute plain language and twenty (20) code groups per minute.

(ii) Written examination element: 7.

(6) At a ship radar station, the holder of this class of license may not supervise or be responsible for the performance of any adjustments or tests during or coincident with the installation, servicing or maintenance of the radar equipment while it is radiating energy unless he has satisfactorily completed a supplementary

examination qualifying him for that duty and received a ship radar endorsement on his license certifying to that fact: *Provided,* That nothing in this subparagraph shall be construed to prevent persons holding licenses not so endorsed from making replacements of fuses or of receiving-type tubes. The supplementary examination shall consist of:

(i) Written examination element: 8.

(c) [Reserved].

(d) *Radiotelegraph third-class operator permit.* Any station except:

(1) Stations transmitting television, or

(2) Any of the various classes of broadcast stations, or

(3) Class I–B coast stations (other than when transmitting manual radiotelegraphy for identification or for testing) at which the power in the antenna of the unmodulated carrier wave is authorized to exceed 250 watts, or

(4) Class II–B or Class III–B coast stations (other than those in Alaska and other than when transmitting manual radiotelegraphy for identification or for testing) at which the power in the antenna of the unmodulated carrier wave is authorized to exceed 250 watts, or

(5) Ship stations or aircraft stations other than those at which the installation is used solely for telephony and at which the power in the antenna of the unmodulated carrier wave is not authorized to exceed 250 watts, or

(6) Ship stations and coast stations open to public correspondence by telegraphy, or

(7) Radiotelegraph stations on board a vessel required by treaty or statute to be equipped with a radio installation, or

(8) Aircraft stations while employing radiotelegraphy:

Provided, That (1) such operator is prohibited from making any adjustments that may result in improper transmitter operation, and (2) the equipment is so designed that the stability of the frequencies of the transmitter is maintained by the transmitter itself within the limits of tolerance specified by the station license, and none of the operations necessary to be performed during the course of normal rendition of the service of the station may cause off-frequency operation or result in any unauthorized radiation, and (3) any needed adjustments of the transmitter that may affect the proper operation of the station are regularly made by or under the immediate supervision and responsibility of a person holding a first- or second-class commercial radio operator license, either radiotelephone or radiotelegraph as may be appropriate for the class of station involved (as determined by the scope of the authority of the respective licenses as set forth in paragraphs (a), (b), (e), and (f) of this section and § 13.62), who shall be responsible for the proper functioning of the station equipment, and (4) in the case of ship radio telephone or aircraft radiotelephone stations when the power in the antenna of the unmodulated carrier wave is authorized to exceed 100 watts, any needed adjustments of the transmitter that may affect the proper operation of the station are made only by or under the immediate supervision and responsibility of an operator holding a first- or second-class radiotelegraph license, who shall be responsible for the proper functioning of the station equipment.

(e) *Radiotelephone first-class operator license.* Any station except:

(1) Stations transmitting telegraphy by any type of the Morse code, or

(2) [Reserved]

(3) At a ship radar station, the holder of this class of license may not supervise or be responsible for the performance of any adjustments or tests during or coincident with the installation, servicing or maintenance of the radar equipment while it is radiating energy unless he has satisfactorily completed a supplementary examination qualifying him for that duty and received a ship radar endorsement on his license certifying to that fact: *Provided,* That nothing in this subparagraph shall be construed to prevent persons holding licenses not so endorsed from making replacements of fuses or of receiving-type tubes. The supplementary examination shall consist of:

(i) Written examination element: 8.

(f) *Radiotelephone second-class operator license.* Any station except:

(1) Stations transmitting telegraphy by any type of the Morse Code, or

(2) Standard broadcast stations, or

(3) International broadcast stations, or

(4) FM broadcast stations, or

(5) Non-commercial educational FM broadcast stations with transmitter power rating in excess of 1 kilowatt, or

(6) Television broadcast stations, or

(7) [Reserved]

(8) At a ship radar station, the holder of this class of license may not supervise or be responsible for the performance of any adjustments or tests during or coincident with the installation, servicing or maintenance of the radar equipment while it is radiating energy unless he has satisfactorily completed a supplementary examination qualifying him for that duty and received a ship radar endorsement on his license certifying to that fact: *Provided,* That nothing in this subparagraph shall be construed to prevent persons holding licenses not so endorsed from making replacements of fuses or of receiving-type tubes. The supplementary examination shall consist of:

(i) Written examination element: 8.

(g) *Radiotelephone third-class operator permit.* Any station except:

(1) Stations transmitting television other than Instructional Television Fixed Service stations, or

(2) Stations transmitting telegraphy by any type of the Morse Code, or

(3) Any of the various classes of broadcast stations, or

(4) Class I–B coast stations at which the power is authorized to exceed 250 watts carrier power or 1,000 watts peak envelope power, or

(5) Class II–B or Class III–B coast stations, other than those in Alaska, at which the power is authorized to exceed 250 watts carrier power or 1,000 watts peak envelope power, or

(6) Ship stations or aircraft stations at which the installation is not used solely for telephony, direct-printing or at which the power is more than 250 watts carrier power or 1,000 watts peak envelope power: *Provided,* That (1) such operator is prohibited from making any adjustments that may result in improper transmitter operation, and (2) the equipment is so designed that the stability of the frequencies of the transmitter is maintained by the transmitter itself within the limits of tolerance specified by the station license, and none of the operations necessary to be performed during the course of normal rendition of the service of the station may cause off-frequency operation or result in any unauthorized radiation, and (3) any needed adjustments of the transmitter that may affect the proper operation of the station are regularly made by or under the immediate supervision and responsibility of a person holding a first- or second-class commercial radio operator license, either radiotelephone or radiotelegraph as may be appropriate for the class of station involved (as determined by the scope of the authority of the respective licenses as set forth in paragraphs (a), (b), (e), and (f) of this section and § 13.62), who shall be responsible for the proper functioning of the station equipment, and (4) in the case of ship radiotelephone or aircraft radiotelephone stations when the power in the antenna of the unmodulated carrier wave is authorized to exceed 100 watts, any needed adjustments of the transmitter that may affect the proper operation of the station are made only by or under the immediate supervision and responsibility of an operator holding a first- or second-class radiotelegraph license, who shall be responsible for the proper functioning of the station equipment.

(h) *Restricted radiotelephone operator permit,* any station except:

(1) Stations transmitting television, or

(2) Stations transmitting telegraphy by any type of the Morse Code, or

(3) Any of the various classes of broadcast stations other than FM translator and booster stations, or

(4) Ship stations licensed to use telephony at which the power is more than 100 watts carrier power or 400 watts peak envelope power, or

(5) Radio stations provided on board vessels for safety purposes pursuant to statute or treaty, or

(6) Coast stations, other than those in Alaska, while employing a frequency below 30 MHz, or

(7) Coast stations at which the power is authorized to exceed 250 watts carrier power or 1,000 watts peak envelope power;

(8) At a ship radar station the holder of this class of license may not supervise or be responsible for the performance of any adjustments or tests during or coincident with the installation, servicing or maintenance of the radar equipment while it is radiating energy: *Provided,* That nothing in this subparagraph shall be construed to prevent any person holding such a license from making replacements of fuses or of receiving type tubes:

Provided, That, with respect to any station which the holder of this class of license may operate, such operator is prohibited from making any adjustments that may result in improper transmitter operation, and the equipment is so designed that the stability of the frequencies of the transmitter is maintained by the transmitter itself within the limits of tolerance specified by the station license, and none of the operations necessary to be performed during the course of normal rendition of the service of the station may cause off-frequency operation or result in any unauthorized radiation, and any needed adjustments of the transmitter that may affect the proper operation of the station are regularly made by or under the

mmediate supervision and responsibility of a person
holding a first- or second-class commercial radio
operator license, either radiotelephone or radiotelegraph, who shall be responsible for the proper functioning of the station equipment.

§ 13.62 Special privileges.

In addition to the operating authority granted
under § 13.61, the following special privileges are
granted the holders of commercial radio operator
licenses:

(a) [Reserved]

(b) The holder of any class of radiotelephone operator's license, whose license authorizes him to operate
a station while transmitting telephony, may operate
the same station when transmitting on the same frequencies, any type of telegraphy under the following
conditions:

(1) When transmitting telegraphy by automatic
means for identification, for testing, or for actuating
an automatic selective signaling device, or

(2) When properly serving as a relay station and
for that purpose retransmitting by automatic means,
solely on frequencies above 50 Mc/s, the signals of a
radiotelegraph station, or

(3) When transmitting telegraphy as an incidental
part of a program intended to be received by the
general public, either directly or through the intermediary of a relay station or stations.

(c) The holder of any class of commercial operator
license may operate AM, FM, or educational FM broadcast stations, except AM stations using directional antenna systems which are required by the station authorizations to maintain ratios of the currents in the
elements of the systems within tolerance which is
less than five percent or relative phases within tolerances which are less than three degrees, under the following conditions:

(1) That adjustments of transmitting equipment by
such operators, except when under the immediate supervision of a radiotelephone first class operator (radiotelephone second class operator for educational FM
stations with transmitter output power of 1000 watts
or less) shall be limited to the following:

(i) Those necessary to turn the transmitter on and
off;

(ii) Those necessary to compensate for voltage fluc-
uations in the primary power supply;

(iii) Those necessary to maintain modulation levels
of the transmitter within prescribed limits;

(iv) Those necessary to effect routine changes in
operating power which are required by the station authorization;

(v) Those necessary to change between nondirectional and directional or between differing radiation
patterns, provided that such changes require only activation of switches and do not involve the manual tuning of the transmitter's final amplifier or antenna
phasor equipment. The switching equipment shall be
so arranged that the failure of any relay in the directional antenna system to activate properly will cause
the emissions of the station to terminate.

(2) The emissions of the station shall be terminated
immediately whenever the transmitting system is observed operating beyond the upper and lower limiting
values of parameters required to be observed and
logged or in any manner inconsistent with the rules
or the station authorization, and the above adjustments are ineffective in correcting the condition of
improper operation, and a first class radiotelephone
operator is not present (radiotelephone second class
operator for educational FM stations with transmitter
output power of 1000 watts or less).

(3) The special operating authority granted in this
section with respect to broadcast stations is subject to
the condition that there shall be in employment at the
station in accordance with Part 73 of this chapter one
or more first-class radiotelephone operators authorized
to make or supervise all adjustments. whose primary
duty shall be to affect and insure the proper functioning of the transmitting system. In the case of a noncommercial educational FM broadcast station with authorized transmitter output power of 1000 watts or less,
a second-class radiotelephone licensed operator may be
employed in lieu of a first-class licensed operator.

*   *   *   *   *   *   *

§ 13.64 Obedience to lawful orders.

All licensed radio operators shall obey and carry
out the lawful orders of the master or person lawfully
in charge of the ship or aircraft on which they are
employed.

§ 13.65 Damage to apparatus.

No licensed radio operator shall willfully damage,
or cause or permit to be damaged, any radio apparatus or installation in any licensed radio station.

§ 13.66 Unnecessary, unidentified, or superfluous com-
munications.

No licensed radio operator shall transmit unnecessary, unidentified, or superfluous radio communications or signals.

§ 13.67 Obscenity, indecency, profanity.

No licensed radio operator or other person shall
transmit communications containing obscene, indecent, or profane words, language, or meaning.

§ 13.68 False signals.

No licensed radio operator shall transmit false or
deceptive signals or communications by radio, or any
call letter or signal which has not been assigned by
proper authority to the radio station he is operating.

§ 13.69 Interference.

No licensed radio operator shall willfully or maliciously interfere with or cause interference to any
radio communication or signal.

§ 13.70 Fraudulent licenses.

No licensed radio operator or other person shall
alter, duplicate, or fraudulently obtain or attempt to
obtain, or assist another to alter, duplicate, or fraudulently obtain or attempt to obtain an operator license.
Nor shall any person use a license issued to another or
a license that he knows to be altered, duplicated, or
fraudulently obtained.

MISCELLANEOUS

§ 13.71 Issue of duplicate or replacement licenses.

(a) If the authorization is of the diploma form, a
properly executed application for duplicate should be
submitted to the office of issue. If the authorization is
of the card form (Restricted Radiotelephone Operator
Permit), a properly executed application for replacement should be submitted to the Federal Communications Commission, Gettysburg, Pa., 17325, EXCEPT for
alien restricted radiotelephone operator permit applica-

157

tions, which must be submitted to Federal Communications Commission, Washington, D.C. 20554. In either case, the application shall embody a statement of the circumstances involved in the loss, multilation, or destruction of the license or permit. If the authorization has been lost, the applicant must state that reasonable search has been made for it, and further, that in the event it be found, either the original or the duplicate (or replacement) will be returned for cancellation. If the authorization is of the diploma form, the applicant should also submit documentary evidence of the service that has been obtained under the original authorization, or a statement embodying that information.

(b) The holder of any license or permit whose name is legally changed may make application for a replacement document to indicate the new legal name by submitting a properly executed application accompanied by the license or permit affected. If the authorization is of the diploma form, the application should be submitted to the office where it was issued. If the authorization is of the card form (Restricted Radiotelephone Operator Permit) it should be submitted to the Federal Communications Commission, Gettysburg, Pa. 17325, except for alien restricted radiotelephone operator permit applications, which must be submitted to Federal Communications Commission, Washington, D.C. 20554.

*   *   *   *   *   *   *

§ 13.72 Exhibiting signed copy of application.

When a duplicate or replacement operator license or permit has been requested, or request has been made for renewal, or a request has been made for an endorsement, higher class license or permit, or verification card, the operator shall exhibit in lieu of the original document a signed copy of the application which has been submitted to the Commission.

§ 13.73 Verification card.

The holder of an operator license or permit of the diploma form (as distinguished from such document of the card form) may, by filing a properly executed application accompanied by his license or permit, obtain a verification card (Form 758–F). This card may be carried on the person of the operator in lieu of the original license or permit when operating any station at which posting of an operator license is not required: *Provided*, That the license is readily accessible within a reasonable time for inspection upon demand by an authorized Government representative.

*   *   *   *   *   *   *

Aviation Red Obstruction Lighting

§ 17.24 Specifications for the lighting of antenna structures up to and including 45.72 meters (150 feet) in height.

Antenna structures up to and including 150 feet in height above ground, which are required to be lighted as a result of notification to the FAA under § 17.7, shall be lighted as follows:

(a) There shall be installed at the top of the tower at least two 116- or 125-watt lamps (A21/TS) enclosed in aviation red obstruction light globes. The intensity of each lamp shall not be less than 32.5 candelas. The two lights shall burn simultaneously from sunset to sunrise and shall be positioned so as to insure unobstructed visibility of at least one of the lights from aircraft at any normal angle of approach. A light sensitive control device or an astronomic dial clock and time

switch may be used to control the obstruction lighting in lieu of manual control. When a light sensitive device is used, it shall be adjusted so that the lights will be turned on when the north sky illuminance on a vertical surface falls to a level of not less than 376.74 lux (35 fc) and turned off when the north sky illuminance on a vertical surface rises to a level of not less than 624.31 lux (58 fc).

§ 17.25 Specifications for the lighting of antenna structures over 45.72 meters (150 feet) up to and including 91.44 meters (300 feet) in height.

(a) Antenna structures over 45.72 meters (150 feet) up to and including 60.96 meters (200 feet) in height above ground, which are required to be lighted as a result of notification to the FAA under § 17.7 and antenna structures over 60.96 meters (200 feet) up to and including 91.44 meters (300 feet) in height above ground, shall be lighted as follows:

(1) There shall be installed at the top of the structure one 300 m/m electric code beacon equipped with two 620- or 700-watt lamps (PS–40 Code Beacon type) both lamps to burn simultaneously, and equipped with aviation red color filters. The steady burning intensity shall not be less than 2,000 candelas (in red.) Where a rod or other construction of not more than 6.10 meters (20 feet) in height and incapable of supporting this beacon is mounted on top of the structure and it is determined that this additional construction does not permit unobstructed visibility of the code beacon from aircraft at any normal angle of approach, there shall be installed two such beacons positioned so as to insure unobstructed visibility of at least one of the beacons from aircraft at any normal angle of approach. The beacons shall be equipped with a flashing mechanism producing not more than 40 flashes per minute nor less than 12 flashes per minute, with a period of darkness equal to approximately one-half of the luminous period.

(2) At the approximate mid point of the overall height of the tower there shall be installed at least two 116- or 125-watt lamps (A21/TS) enclosed in aviation red obstruction light globes. The intensity of each lamp shall not be less than 32.5 candelas. Each light shall be mounted so as to insure unobstructed visibility of at least one light at each level from aircraft at any normal angle of approach.

(3) All lights shall burn continuously or shall be controlled by a light sensitive device adjusted so that the lights will be turned on when the north sky illuminance on a vertical surface falls to a level of not less than 376.74 lux (35 fc) and turned off when the north sky illuminance on a vertical surface rises to a level of not less than 624.31 lux (58 fc).

§ 17.47 Inspection of tower lights and associated control equipment.

The licensee of any radio station which has an antenna structure requiring illumination pursuant to the provisions of section 303(q) of the Communications Act of 1934, as amended, as outlined elsewhere in this part:

(a) (1) Shall make an observation of the tower lights at least once each 24 hours either visually or by observing an automatic properly maintained indicator designed to register any failure of such lights, to insure that all such lights are functioning properly as required; or alternatively,

(2) Shall provide and properly maintain an auto-

matic alarm system designed to detect any failure of such lights and to provide indication of such failure to the licensee.

(b) Shall inspect at intervals not to exceed 3 months all automatic or mechanical control devices, indicators, and alarm systems associated with the tower lighting to insure that such apparatus is functioning properly.

§ 17.48 Notification of extinguishment or improper functioning of lights.

The licensee of any radio station which has an antenna structure requiring illumination pursuant to the provisions of section 303(q) of the Communications Act of 1934, as amended, as outlined elsewhere in this part:

(a) Shall report immediately by telephone or telegraph to the nearest Flight Service Station or office of the Federal Aviation Administration any observed or otherwise known extinguishment or improper functioning of any top steady burning light or any flashing obstruction light, regardless of its position on the antenna structure, not corrected within 30 minutes. Such reports shall set forth the condition of the light or lights, the circumstances which caused the failure, and the probable date for restoration of service. Further notification by telephone or telegraph shall be given immediately upon resumption of normal operation of the light or lights.

(b) An extinguishment or improper functioning of a steady burning side intermediate light or lights, shall be corrected as soon as possible, but notification to the FAA of such extinguishment or improper functioning is not required.

§ 17.49 Recording of tower light inspections in the station record.

The licensee of any radio station which has an antenna structure requiring illumination shall make the following entries in the station record of the inspections required by § 17.47.

(a) The time the tower lights are turned on and off each day if manually controlled.

(b) The time the daily check of proper operation of the tower lights was made, if automatic alarm system is not provided.

(c) In the event of any observed or otherwise known extinguishment or improper functioning of a tower light:

(1) Nature of such extinguishment or improper functioning.

(2) Date and time the extinguishment or improper functioning was observed, or otherwise noted.

(3) Date, time, and nature of the adjustments, repairs, or replacements made.

(4) Identification of Flight Service Station (Federal Aviation Administration) notified of the extinguishment of improper functioning of any code or rotating beacon light or top light not corrected within 30 minutes, and the date and time such notice was given.

(5) Date and time notice was given to the Flight Service Station (Federal Aviation Administration) that the required illumination was resumed.

(d) Upon completion of the periodic inspection required at least once each 3 months:

(1) The date of the inspection and the condition of all tower lights and associated tower lighting control devices, indicators and alarm systems.

(2) Any adjustments, replacements, or repairs made to insure compliance with the lighting requirements and the date such adjustments, replacements or repairs were made.

*　*　*　*　*　*　*

§ 81.179 Message charges.

(a) (1) No charge shall be made for the service of any public coast station unless effective tariffs applicable to such service are on file with the Commission, pursuant to the requirements of Section 203 of the Communications Act and Part 61 of this chapter.

(2) No charge shall be made for the service of any station subject to this part, other than a public coast or Alaska public fixed station, except as provided by and in accordance with § 81.352.

(b) No charge shall be made by any station in the maritime mobile service of the United States for the transmission of distress messages and replies thereto in connection with situations involving the safey of life and property at sea.

(c) No charge shall be made by any station in the maritime mobile service of the United States for the transmission, receipt, or relay of the information concerning dangers to navigation designated in § 83.303 (b) of this chapter, originating on a ship of the United States or of a foreign country.

*　*　*　*　*　*　*

§ 81.302 Points of communication.

(a) Subject to the conditions and limitations imposed by the terms of the particular coast station license or by the applicable provisions of this part with respect to the use of particular radiochannels, public coast stations using telephony are authorized to communicate:

(1) With any ship station or aircraft station operating in the maritime mobile service for the transmission or reception of safety communication;

(2) With any land station for the purpose of facilitating the transmission or reception of safety communication to or from a ship or aircraft station;

*　*　*　*　*　*　*

§ 83.6 Operational.
*　*　*　*　*　*　*

(f) *Calling.* Transmission from a station solely to secure the attention of another station, or other stations, for a particular purpose.

(g) *Working.* Radiocommunication carried on, for a purpose other than calling, by any station or stations using telegraphy, telephony, or facsimile.

*　*　*　*　*　*　*

§ 83.165 Posting of operator authorization.

(a) Except as provided in paragraph (b) of this section, when an operator is required for the operation of a station subject to this part, the original authorization of each such operator while he is employed or designated as radio operator of the station shall be posted in a conspicuous place at the principal location on board ship at which the station is operated.

(b) An operator who holds a Restricted Radiotelephone Operator Permit or a valid license verification card (FCC Form 758–F) attesting to the existence of a commercial radio operator license of the diploma type, may, in lieu of posting, have such permit or verification card in his personal possession immediately available for inspection upon request by a Commission representative when operating the following:

(1) A station which is not required to be installed on the vessel by reason of statute or treaty to which the United States is a party;

(2) Any class of ship station when the operator is on board solely for the purpose of servicing the radio equipment;

(3) A station of a portable nature.

*   *   *   *   *   *   *

§ 83.201 Watch required during silence periods.

(a) All ship stations employing telegraphy and normally keeping watch on frequencies in the authorized bands between 405 and 535 kHz shall, during their hours of service, take the necessary measures to insure an efficient watch by a duly licensed radiotelegraph operator on the international distress frequency 500 kHz for three minutes twice each hour, beginning at x h. 15 and x h. 45, Greenwich mean time (GMT). For this purpose, either headphones or a loudspeaker may be used, on condition that use of the loudspeaker is no less effective than use of headphones. While maintaining this watch, the operator shall not use or operate any radio equipment (such as, for examples, broadcast receivers, or amateur transmitters or receivers) not actually required for maritime mobile service.

(b) Except for stations on board vessels required by law to be fitted with radiotelegraph equipment, each ship station (in addition to those ship stations specified in paragraph (a) of this section) licensed to transmit by telephony on one or more frequencies within the band 1605 to 3500 kHz shall, during its hours of service for telephony in this band, and as far as possible, maintain an efficient watch for the reception of A3 and A3H emissions on the authorized carrier frequency 2182 kHz, whenever such station is not being used for transmission on that frequency or for communication on other frequencies in this band. The ship station shall, during its hours of service for radiotelephony, maintain such watch at least twice each hour during the silence period x h:00–x h:03 and x h:30–x h:33 Greenwich mean time. Except for messages of distress, urgency and vital navigational warnings, ship stations shall not transmit on 2182 kHz during the silence periods.

§ 83.224 Watch on 156.8 MHz.

Each ship station, or if more than one maritime mobile station is being operated from a vessel then at least one station, licensed to transmit by telephony on one or more frequencies within the band 156–162 MHz, shall, during its hours of service for radiotelephony in this band, maintain an efficient watch for the reception of F3 emissions on the frequency 156.8 MHz whenever such station is not being used for exchanging communications; Provided, however, that the watch is not required:

(a) When ship stations are operating only with single or dual channel bridge-to-bridge radio equipment

pursuant to Section 83.106(a)(5) of the rules; or

(b) For tug boats, or other towing vessels, of less than 1600 gross tons when operating in a U.S. Coast Guard designated Vessel Traffic Service (VTS) area and maintaining a listening watch on the specified VTS frequency; and provided further that a watch is also maintained on a frequency assigned to a limited coast station as directed by the commercial organization to which the vessel is affiliated.

*   *   *   *   *   *   *

§ 83.234 Distress signals.

(a) The international radiotelegraph distress signal consists of the group "three dots, three dashes, three dots" (. . . — — — . . .), symbolized herein by $\overline{SOS}$, transmitted as a single signal in which the dashes are slightly prolonged so as to be distinguished clearly from the dots.

(b) The international radiotelephone distress signal consists of the word MAYDAY, pronounced as the French expression "m'aider".

(c) These distress signals indicate that a mobile station is threatened by grave and imminent danger and requests immediate assistance.

§ 83.235 Distress calls.

(a) The distress call sent by radiotelegraphy consists of:

(1) The distress signal $\overline{SOS}$, sent three times;

(2) The word DE;

(3) The call sign of the mobile station in distress, sent three times.

(b) The distress call sent by radiotelephony consists of:

(1) The distress signal MAYDAY spoken three times;

(2) The words THIS IS;

(3) The call sign (or name, if no call sign assigned) of the mobile station in distress, spoken three times.

(c) The distress call shall have absolute priority over all other transmissions. All stations which hear it shall immediately cease any transmission capable of interfering with the distress traffic and shall continue to listen on the frequency used for the emission of the distress call. This call shall not be addressed to a particular station and acknowledgment of receipt shall not be given before the distress message which follows it is sent.

§ 83.236 Distress messages.

(a) The radiotelegraph distress message consists of:

(1) The distress signal $\overline{SOS}$;

(2) The name of the mobile station in distress;

(3) Particulars of its position;

(4) The nature of the distress;

(5) The kind of assistance desired;

(6) Any other information which might facilitate rescue.

(b) The radiotelephone distress message consists of:

(1) The distress signal MAYDAY;

(2) The name of the mobile station in distress;

(3) Particulars of its position;

(4) The nature of the distress;

(5) The kind of assistance desired;

(6) Any other information which might facilitate rescue (for example, the length, color, and type of vessel; number of persons on board, etc.).

(c) As a general rule, a ship shall signal its position

in latitude and longitude (Greenwich), using figures
for the degrees and minutes, together with one of the
words NORTH or SOUTH and one of the words EAST
or WEST. In radiotelegraphy, the signal . _ . _ . _
shall be used to separate the degrees from the minutes.
When practicable, the true bearing and distance in
nautical miles from a known geographical position
may be given.

*   *   *   *   *   *   *

§83.238 Radiotelephone distress call and message transmission procedure.

(a) The radiotelephone distress procedure shall consist of:

(1) The radiotelephone alarm signal (whenever possible);

(2) The distress call;

(3) The distress message.

(b) The radiotelephone distress transmissions shall
be made slowly and distinctly, each word being clearly
pronounced to facilitate transcription.

(c) After the transmission by radiotelephony of its
distress message, the mobile station may be requested
to transmit suitable signals followed by its call sign or
name, to permit direction-finding stations to determine
its position. This request may be repeated at frequent
intervals if necessary.

(d) The distress message, preceded by the distress
call, shall be repeated at intervals until an answer is
received. This repetition shall be preceded by the radiotelephone alarm signal whenever possible.

(e) When the mobile station in distress receives no
answer to a distress message transmitted on the distress frequency, the message may be repeated on any
other available frequency on which attention might be
attracted.

§83.239 Acknowledgment of receipt of distress message.

(a) Stations of the maritime mobile service which
receive a distress message from a mobile station which
is, beyond any possible doubt, in their vicinity, shall
immediately acknowledge receipt. However, in areas
where reliable communication with one or more coast
stations are practicable, ship stations may defer this
acknowledgment for a short interval so that a coast
station may acknowledge receipt.

(b) Stations of the maritime mobile service which
receive a distress message from a mobile station which,
beyond any possible doubt, is not in their vicinity, shall
allow a short interval of time to elapse before acknowledging receipt of the message, in order to permit stations nearer to the mobile station in distress to acknowledge receipt without interference.

§83.240 Form of acknowledgment.

(a) The acknowledgement of receipt of a distress
message is transmitted, when radiotelegraphy is
used, in the following form:

(1) The call sign of the station sending the distress
message, sent three times;

(2) The word DE;

(3) The call sign of the station acknowledging receipt, sent three times;

(4) The group RRR;

(5) The distress signal SOS

(b) The acknowledgment of receipt of a distress
message is transmitted, when radiotelephony is used,
in the following form:

(1) The call sign or other identification of the station sending the distress message, spoken three times;

(2) The words THIS IS;

(3) The call sign or other identification of the station acknowledging receipt, spoken three times;

(4) The word RECEIVED;

(5) The distress signal MAYDAY.

§83.241 Information furnished by acknowledging station.

(a) Every mobile station which acknowledges receipt of a distress message shall, on the order of the
master or person responsible for the ship, aircraft, or
other vehicle carrying such mobile station, transmit as
soon as possible the following information in the order
shown:

(1) Its name;

(2) Its position, in the f o r m prescribed in
§83.236(c);

(3) The speed at which it is proceeding towards,
and the approximate time it will take to reach, the
mobile station in distress.

(b) Before sending this message, the station shall
ensure that it will not interfere with the emissions of
other stations better situated to render immediate assistance to the station in distress.

§83.242 Transmission of distress message by a station not itself in distress.

(a) A mobile station or a land station which learns
that a mobile station is in distress shall transmit a
distress message in any of the following cases:

(1) When the station in distress is not itself in a
position to transmit the distress message;

(2) When the master or person responsible for the
ship, aircraft, or other vehicle not in distress, or the
person responsible for the land station, considers that
further help is necessary;

(3) When, although not in a position to render assistance, it has heard a distress message which has not
been acknowledged. When a mobile station transmits
a distress message under these conditions, it shall
take all necessary steps to notify the authorities who
may be able to render assistance.

(b) The transmission of a distress message under
the conditions prescribed in paragraph (a) of this
section shall be made on either or all of the international distress frequencies (500 kHz radiotelegraph;
2182 kHz or 156.8 MHz radiotelephone) or on any
other available frequency on which attention might
be attracted.

(c) The transmission of the distress message shall
always be preceded by the call indicated below, which
shall itself be preceded whenever possible by the radiotelegraph or radiotelephone alarm signal. This call
consists of:

(1) When radiotelegraphy is used:

(i) the signal $\overline{DDD}$ SOS SOS SOS $\overline{DDD}$;

(ii) The word DE;

(iii) The call sign of the transmitting station, sent
three times.

(2) When radiotelephony is used:

(i) The signal MAYDAY RELAY, spoken three
times;

(ii) The words THIS IS;

(iii) The call sign or other identification of the
trnsmitting station, spoken three times.

(d) When the radiotelegraph alarm signal is used,

an interval of two minutes shall be allowed, whenever this is considered necessary, before the transmission of the call mentioned in subparagraph (c)(1) of this section.

§ 83.243 Control of distress traffic.

(a) Distress traffic consists of all messages relating to the immediate assistance required by the mobile station in distress. In distress traffic, the distress signal shall be sent before the call and at the beginning of the preamble of any radiotelegram.

(b) The control of distress traffic is the responsibility of the mobile station in distress or of the station which, pursuant to § 83.242(a), has sent the distress message. These stations may, however, delegate the control of the distress traffic to another station.

(c) The station in distress or the station in control of distress traffic may impose silence either on all stations of the mobile service in the area or on any station which interferes with the distress traffic. It shall address these instructions "to all stations" or to one station only, according to circumstances. In either case, it shall use:

(1) In radiotelegraphy, the abbreviation QRT, followed by the distress signal SOS. The use of the signal QRT SOS shall be reserved for the mobile station in distress and for the station controlling distress traffic;

(2) In radiotelephony, the signal SEELONCE MAYDAY. The use of this signal shall be reserved for the mobile station in distress and for the station controlling distress traffic.

(d) If it is believed to be essential, any station of the mobile service near the ship, aircraft, or other vehicle in distress, may also impose silence. It shall use for this purpose:

(1) In radiotelegraphy, the abbreviation QRT, followed by the word DISTRESS and its own call sign;

(2) In radiotelephony, the word SEELONCE, followed by the word DISTRESS and its own call sign or other identification.

*   *   *   *   *   *   *

§ 83.247 Urgency signals.

(a) The urgency signal indicates that the calling station has a very urgent message to transmit concerning the safety of a ship, aircraft, or other vehicle, or the safety of a person. The urgency signal shall be sent only on the authority of the master or person responsible for the mobile station.

*   *   *   *   *   *   *

(c) In radiotelephony, the urgency signal consists of the word PAN, spoken three times and transmitted before the call.

*   *   *   *   *   *   *

§ 83.249 Safety signals.

(a) The safety signal indicates that the station is about to transmit a message concerning the safety of navigation or giving important meteorological warnings.

(b) In radiotelegraphy, the safety signal consists of three repetitions of the group TTT, sent with the individual letters of each group, and the successive groups clearly separated from each other. It shall be sent before the call.

(c) In radiotelephony, the safety signal consists of the word SECURITY, spoken three times and transmitted before the call.

(d) The safety signal and call shall be sent on one of the international distress frequencies (500 kHz radiotelegraph; 2182 kHz or 156.8 MHz radiotelephone). However, stations which cannot transmit on a distress frequency may use any other available frequency on which attention might be attracted.

*   *   *   *   *   *   *

§ 83.352 Frequencies for use in distress and search and rescue operations.

(a) The frequency 2182 kHz is the international distress frequency for radiotelephony; it shall be used for this purpose by ship ,aircraft, and survival craft stations operating in the authorized bands between 1605 and 4000 kHz when requesting assistance from the maritime services. The frequency 156.8 MHz is the international distress, safety and calling frequency for radiotelephony for stations of the maritime mobile service when using frequencies in the authorized bands between 156 and 174 MHz.

(b) The frequency 121.5 MHz (using class A2 emission) is available for assignment to survival craft stations for radiobeacon purposes. The frequency 121.5 MHz (using class A9 emission) is available for assignment to stations for facilitating search and rescue operations. The frequencies 121.5 MHz and 123.1 MHz (class A3 emission) are available for assignment to ship stations for scene of action search and rescue operations between ships and aircraft. The frequency 121.5 MHz will be assigned to ship stations for search and rescue operations only if the frequency 123.1 MHz is also authorized. Communication in support of search and rescue operations is permitted on the frequency 121.5 MHz only when communications on the frequency 123.1 MHz or other VHF frequencies is not practicable. Ships and aircraft engaged in such communications on 121.5 MHz should shift to 123.1 MHz as soon as possible.

(c) The frequency 243 MHz (class A9 emission only) is available to EPIRB stations for facilitating search and rescue operations.

§ 83.353 Frequencies for calling.

The frequency 2182 kHz is the international general calling frequency for radiotelephony; it may be used for this purpose by ship and aircraft stations operating in the maritime mobile service in the bands between 1605 and 4000 kHz. The frequency 156.8 MHz is the international distress, safety and calling frequency for radiotelephony for stations of the maritime mobile service when using frequencies in the authorized bands between 156 and 174 MHz. In addition, these frequencies may be used for transmission of:

(a) The international urgency signal, and very urgent messages (preceded by this signal) concerning the safety of a ship, aircraft, or other vehicle, or the safety of some person on board or within sight of such ship, aircraft, or vehicle.

(b) The international safety signal, and messages (preceded by this signal) concerning the safety of navigation or giving important meteorological warnings; however, safety messages shall be transmitted, when practicable, on a working frequency after a preliminary anouncement on 2182 kHz.

(c) Brief radio operating signals.

(d) Brief test signals in accordance with the provisions of § 83.365, as may be necessary to determine whether the radio transmitting equipment of the station is in good working condition on this frequency.

*　　*　　*　　*　　*　　*　　*

§ 83.365 Procedure in testing.

(a) Ship stations must use every precaution to insure that, when conducting operational transmitter tests, the emissions of the station will not cause harmful interference. Radiation must be reduced to the lowest practicable value and if feasible shall be entirely suppressed. When radiation is necessary or unavoidable, the testing procedure described below shall be followed:

(1) The licensed radio operator or other person responsible for operation of the transmitting apparatus shall ascertain by careful listening that the test emissions will not be likely to interfere with transmissions in progress; if they are likely to interfere with the working of a coast or aeronautical station in the vicinity of the ship station, the consent of the former station(s) must be obtained before the test emissions occur; (see required procedures in subparagraphs (2) and (3) of this paragraph following);

(2) The applicable identification of the testing station, followed by the word "test" shall be announced on the radio-channel being used for the test, as a warning that test emissions are about to be made on that frequency;

(3) If, as a result of the announcement prescribed in subparagraph (2) of this paragraph, any station transmits by voice the word "wait", testing shall be suspended. When, after an appropriate interval of time such announcement is repeated and no response is observed, and careful listening indicates that harmful interference should not be caused, the operator shall, if further testing is necessary, proceed as set forth in subparagraphs (4) and (5) of this paragraph:

(4) Testing of transmitters shall, insofar as practicable, be confined to working frequencies without two way communications; however, 2182 kHz and 156.8 MHz may be used to contact other ship or coast stations when signal reports are necessary. Short tests, by vessels which continue to rely upon the use of DSB equipment for distress and safety purposes, are permitted on 2182 kHz to evaluate the compatibility of that equipment with an SSB emission A3J system. U.S. Coast Guard stations may be contacted on 2182 kHz for test purposes only when tests are being conducted during inspections by Commission representatives or when qualified radio technicians are installing equipment or correcting deficiencies in the station radiotelephone equipment. In these cases the test shall be identified as "FCC" or "technical" and logged accordingly;

(5) When further testing is necessary beyond the two "test" announcements specified in subparagraphs (2) and (3) of this paragraph, the operator shall announce the word "testing" followed in the case of a voice transmission test by the count "1, 2, 3, 4, • • • etc." or by test phrases or sentences not in conflict with normal operating signals. The test signals in either case shall have a duration not exceeding 10 seconds. At the conclusion of the test, there shall be voice announcement of the official call sign of the testing station. This test transmission shall not be repeated until a period of at least 1 minute has elapsed; on the frequency 2182 kHz or 156.8 MHz a period of at least 5 minutes shall elapse before the test transmission is repeated.

(b) When testing is conducted on any frequency within the bands 2173.5 to 2190.5 kHz, 156.75 to 156.85, 480 to 510 kHz (survival craft transmitters only), or 8362 to 8366 kHz (survival craft transmitters only), no test transmissions shall occur which are likely to actuate any automatic alarm receiver within range. Survival craft stations shall not be tested on the frequency 500 kHz during the 500 kHz silence periods.

§ 83.366 General radiotelephone operating procedure.

(a) *Calling coast stations.* (1) Use by ship stations of the frequency 2182 kHz for calling coast stations, and for replying to calls from coast stations, is authorized; however, whenever practicable such calls and replies shall be made on the appropriate ship-shore working frequency.

(2) Use by ship stations and marine utility stations on board ship of the frequency 156.8 MHz for calling coast stations and marine utility stations on shore, and for replying to calls from such stations, is authorized; however, whenever practicable such calls and replies shall be made on the appropriate ship-shore working frequency.

(b) *Calling ship stations.* (1) Except when other operating procedure is used to expedite safety communication, ship stations, before transmitting on the intership working frequencies 2003, 2142, 2638, 2738, or 2830 kHz, shall first establish communication with other ship stations by call and reply on 2182 kHz: *Provided,* That calls may be initiated on an intership working frequency when it is known that the called vessel maintains a simultaneous watch on such working frequency and on 2182 kHz.

(2) Except when other operating procedure is used to expedite safety communication, the frequency 156.8 MHz shall be used for call and reply by ship stations and marine utility stations on board ship before establishing communication on either of the intership working frequencies 156.3 or 156.4 MHz.

(c) *Change to working frequency.* After establishing communication with another station by call and reply on 2182 kHz or 156.8 MHz, stations on board ship shall change to an authorized working frequency for the transmission of messages which, under the provisions of this subpart, cannot be transmitted on the respective calling frequencies.

(d) *Authorized use of 2003, 2142, 2638, 2738, and 2830 kHz.* The intership working frequencies 2003, 2142, 2638, 2738, and 2830 kHz shall be used for transmissions by ship stations in accordance with the provisions of §§ 83.176, 83.177, and 83.358.

(e) *Simplex operation only.* All transmission on 2003, 2142, 2638, 2738, and 2830 kHz by two or more stations, engaged in any one exchange of signals or communications, shall take place on only one of these frequencies, i.e., the stations involved shall transmit and receive on the same frequency: *Provided,* That this requirement is waived in the event of emergency when by reason of interference or limitation of equipment single frequency operation cannot be used.

(f) *Limitation on duration of calling.* Calling a particular station shall not continue for more than 30 seconds in each instance. If the called station is not heard to reply, that station shall not again be called until after an interval of 2 minutes. When a station called does not reply to a call sent three times at intervals of 2 minutes, the calling shall cease and shall not be renewed until after an interval of 15 minutes; however, if there is no reason to believe that harmful interference will be caused to other communications in progress, the call sent three times at intervals of 2 minutes may be repeated after a pause of not less than 3 minutes. In event of an emergency involving safety, the provisions of this paragraph shall not apply.

(g) *Limitation on duration of working.* Any one exchange of communications between any two ship stations on 2003, 2142, 2638, 2738, or 2830 kHz, or between a ship station and a limited coast station on 2738 or 2830 kHz, shall not exceed 3 minutes in duration after the two stations have established contact by calling and answering. Subsequent to such exchange of communications, the same two stations shall not again use 2003, 2142, 2638, 2738, or 2830 kHz for communication with each other until 10 minutes have elapsed: *Provided,* That this provision shall in no way limit or delay the transmission of communications concerning the safety of life or property.

(h) *Transmission limitation on 2182 kHz and 156.8 MHz.* Except in cases of distress, urgency, or safety, any one exchange of communications on 2182 kHz and 156.8 MHz shall be kept to a minimum and shall not exceed one minute.

(i) *Limitation on commercial communication.* On frequencies in the band 156–162 MHz, the exchange of commercial communication shall be limited to the minimum practicable transmission time. In the conduct of ship-shore communication, other than distress, stations on board ship shall comply with instructions given by the limited coast station or marine utility station on shore with which they are communicating, in all matters relative to operating practices and procedures and to the suspension of transmission in order to minimize interference.

(j) *2182 kHz silence periods.* Transmission by ship or survival craft stations is prohibited on any frequency (including 2182 kHz) within the band 2173.5 to 2190.5 kHz during each 2182 kHz silence period, i.e., for three minutes twice each hour beginning at x h:00 and x h:30, Greenwich mean time: *Provided, however,* That this provision is not applicable to the transmission of distress, urgency and vital navigational warning, or to messages preceded by one of these signals.

Citizens Band (CB) Radio Service Rules

§95.401
CITIZENS BAND (CB) RADIO
SERVICE RULES
GENERAL PROVISIONS

CB Rule 1   What is the Citizens Band (CB) Radio Service?

The CB Radio Service is a private, two-way, short-distance voice communications service for personal or business activities. The CB Radio Service may also be used for voice paging.

CB Rule 2   How do I use these rules?

(a) Read and obey the rules. See CB Rule 37 for the penalties for violations of these rules.

(b) Where the rules use the word "you," "you" means an applicant, a licensee or an individual holding a valid temporary permit, where appropriate.

(c) Where the rules use the word "person," the rules are concerned with any person, including an individual, a corporation, a partnership, or an association.

HOW TO APPLY FOR A CB LICENSE

CB Rule 3   Do I need a license?

Before operating a CB transmitter, you must have authority from the FCC, as follows:

AN INDIVIDUAL MUST:
Get a CB license from the FCC; *OR* Have a properly filled-out temporary permit (FCC Form 555-B); *OR* Qualify to operate a CB transmitter under the authority of another person's license.
AN ASSOCIATION, PARTNERSHIP, CORPORATION, OR GOVERNMENTAL UNIT MUST:
Get a CB license from the FCC; *OR* Request, receive, and comply with a special temporary authority or other special authorization from the FCC.

CB Rule 4   Am I eligible to get a CB license?

(a) You are eligible for a CB license if—	
you are:	AND you are not:
An individual, and you are eighteen years old or older;	a foreign government
A partnership, and each partner is eighteen years old or older;	OR
A corporation;	a representative of
An association;	a foreign government
A state, territorial or local governmental unit; or	OR
Other legal entity	a federal government agency

165

(b) You must not have more than one CB license at any one time.

(c) Any agency operating under the authority of an eligible governmental unit, including an authorized Civil Defense agency, is also eligible for a CB license.

(d) A subsidiary or division of a corporation is not eligible for its own CB license unless the subsidiary or division is separately incorporated.

CB Rule 5   How do I apply for a CB license?

(a) You apply for a CB license by filling out an application (FCC Form 505) and sending it to the FCC, Gettysburg, Pa. 17326.

(b) You can get applications from the FCC; Washington, D.C. 20554 or from any FCC field office. (A list of FCC field offices is contained in CB Rule 45.) Many CB equipment dealers also have application forms.

(c) If you have questions about your application, you should write to the Personal Radio Division, FCC, Washington, D.C. 20554.

(d) If your application is not completely filled out, if you do not make the necessary certifications, or you do not include all necessary information with your application, the FCC may return your application.

(e) A Canadian General Radio Service licensee may apply for permission to operate his or her station in the United States by filling out an application (FCC Form 410-B) and sending it to the FCC, Gettysburg, Pa. 17325.

CB Rule 6   May I operate my CB station while my application is being processed?

(a) If you are an individual, you may operate your CB transmitter after you have mailed your CB license application to the FCC, if—

(1) You fill out a temporary permit application (FCC Form 555-B), and

(2) You *keep* this form with your station records. The completed form *is* your temporary permit.

(b) A CB temporary permit is valid for 60 days after you mail your CB license application to the FCC.

CB Rule 7   We are not an individual. How do we apply for temporary privileges?

(a) Only an individual applicant may use a temporary CB permit.

(b) A partnership, corporation, association, joint-stock company, trust or governmental unit may operate a CB transmitter while its application for a new CB license is pending only if it has obtained special temporary authority from the FCC. A written request for special temporary authority, including a justification for the request, may be submitted to the Personal Radio Division, FCC, Washington, D.C. 20554.

CB Rule 8   How do I renew or modify my CB license?

(a) You renew or modify your license in the same way that you apply for a new CB license. You should allow at least sixty days for the FCC to act on your application.

(b) If you send your application before your license expires, you may continue to operate under that license until the FCC acts on your application. You do not need a temporary permit, but you should keep a copy of the application you send to the FCC.

(c) You must stop transmitting as soon as your license expires, unless you have already sent your renewal application to the FCC. You may not begin transmitting again until you have received a new license from the FCC.

CB Rule 9   How does a corporation holding a CB license apply for consent to transfer control of the corporation?

If a corporation holds a CB license, it must obtain written permission from the FCC before it transfers control of the corporation. A request for this consent must be made on FCC Form 703, and must be sent to the FCC; Washington, D.C. 20554.

CB Rule 10   What address do I put on my application?

(a) You must include your current complete mailing address in the United States and station address on your CB license application.

(b) A Canadian General Radio Service licensee may supply a Canadian address, if he or she is applying for permission to operate a General Radio Service station in the United States. A Canadian General Radio Service licensee applies for permission to operate a General Radio Service station in the United States on FCC Form 410-B.

CB Rule 11   How do I sign my CB license application?

(a) If you are an individual, you must sign your own application personally.

(b) If you are not an individual, you must sign your application as follows:

(c) If the FCC requires you to submit additional information, you must sign it in the same way you signed your application.

(d) If you willfully make a false statement on your application, you may be punished by fine, imprisonment and revocation of your station license.

Applicant	Signature
Partnership	One of the partners.
Corporation	Officer.
Association	Member who is an officer.
Governmental Unit	Appropriate elected or appointed official.

CB Rule 12   How long is my license term?

Your CB license term is usually five years from the date the FCC first issued or renewed it. The expiration date is printed on the license.

CB Rule 13   What kind of operation does my license allow?

(a) You must obey all the conditions and terms of your license.

(b) You may operate your CB station from your car, your house, or any other fixed location. (The FCC licenses all CB stations as mobile stations.)

(c) Your CB license allows you to operate with up to 25 transmitters. To use more than 25 transmitters, you must request and receive written permission from the Personal Radio Division, FCC, Washington, D.C. 20554. Attach a letter to your application explaining why you need more than 25 transmitters and how you will control the operation of the transmitters.

CB Rule 14   What must I do if my name or address changes?

(a) If your name, station address, or mailing address changes, you must inform the FCC, Gettysburg, Pa. 17326. Your notice must include the name and address as it appears on your license, the new name or new address, and your call sign. You must keep a copy of this notice in your station records. (Your notice may be in letter form. Your CB license may have a form attached to it which you can also use for this purpose.)

(b) If you hold a CB license, and then incorporate, form a new partnership or form a new association, you must apply for a new CB license.

CB Rule 15   May I transfer my CB license to another person?

(a) You must not let anyone who is not listed in CB Rule 26 operate under your license. You cannot transfer, assign, sell, or give your CB license or its operating authority to another person.

(b) If you sell or give your CB transmitter to another person, you must not transfer your CB license with the transmitter. The new owner of the CB transmitter must obtain a CB license or other authority from the FCC in his or her own name or qualify to operate under CB Rule 26 before he or she can operate the transmitter.

CB Rule 16   Are there any special restrictions on the location of my CB station?

(a) If your CB station will be constructed on land of environmental or historical importance (such as a location significant in American history, architecture or culture), you may be required to provide additional information with your license application and to comply with §1.1305-1.1319 of the FCC's Rules.

(b) If your CB station is located on land controlled by the Department of Defense, you may be required to comply with additional regulations imposed by the commanding officer of the installation.

HOW TO OPERATE A CB STATION

CB Rule 17   On what channels may I operate?

(a) You may transmit on only the following channels (frequencies).

Channel:	Frequency (megahertz)	Channel:	Frequency (megahertz)
1	26.965	20	27.205
2	26.975	21	27.215
3	26.985	22	27.225
4	27.005	23	27.255
5	27.015	24	27.235
6	27.025	25	27.245
7	27.035	26	27.265
8	27.055	27	27.275
		28	27.285
9	27.065	29	27.295
		30	27.305
10	27.075	31	27.315
11	27.085	32	27.325
12	27.105	33	27.335
13	27.115	34	27.345
14	27.125	35	27.355
15	27.135	36	27.365
16	27.155	37	27.375
17	27.165	38	27.385
18	27.175	39	27.395
19	27.185	40	27.405

(b) CHANNEL 9 MAY BE USED ONLY FOR EMERGENCY COMMUNICATIONS OR FOR TRAVELER ASSISTANCE.

(c) YOU MUST, AT ALL TIMES AND ON ALL CHANNELS, GIVE PRIORITY TO EMERGENCY COMMUNICATIONS.

(d) You may use any channel for emergency communications or for traveler assistance.

(e) You must share each channel with other users.

(f) The FCC will not assign any channel for the private or exclusive use of any particular CB station or group of stations.

(g) The FCC will not assign any channel for the private or exclusive use of CB stations transmitting single sideband or AM.

CB Rule 18   How high may I put my antenna?

(a) If your antenna is installed at a fixed location, the antenna structure (whether receiving, transmitting or both) must comply with *either one* of the following:

(1) The highest point must not be more than 6.10 meters (20 feet) higher than the highest point of the building or tree on which it is mounted; or

(2) The highest point must not be more than 18.3 meters (60 feet) above the ground.

(b) If your CB station is located near an airport, and if your antenna structure is more than 6.1 meters (20 feet) high, you may have to obey additional restrictions. The highest point of your antenna must not exceed one meter above the airport elevation for every hundred meters of distance from the nearest point of the nearest airport runway. Differences in ground elevation between your antenna and the airport runway may complicate this formula. If your CB station is near an airport, you may contact the FCC for a worksheet to help you figure the maximum allowable height for your antenna. Consult Part 17 of the FCC's Rules for more information.

> **WARNING: INSTALLATION AND REMOVAL OF CITIZENS BAND BASE STATION ANTENNAS NEAR POWERLINES IS DANGEROUS. FOR YOUR SAFETY, FOLLOW THE INSTALLATION DIRECTIONS INCLUDED WITH YOUR ANTENNA.**

CB Rule 19   What equipment may I use at my CB station?

(a) You must use an FCC type-accepted CB transmitter at your CB station. You can identify an FCC type-accepted transmitter by the type-acceptance label placed on it by the manufacturer. You may examine a list of type-accepted equipment at any FCC Field Office or at FCC Headquarters.

(b) You must not make, or have made, *any* internal modification to a type-accepted CB transmitter. Any internal modification to a type-accepted CB transmitter cancels the type-acceptance.

(c) You must have all internal repairs or internal adjustments to your transmitter made by, or under the direct supervision of, a licensed first- or second-class radiotele-

phone commercial operator. (See CB Rule 41.)

CB Rule 20   How much power may I use?

(a) Your CB transmitter power output must not exceed the following values under any conditions:

AM (A3) ...... 4 watts (carrier power)
SSB (A3J) .... 12 watts (peak envelope power)

(b) If you need more information about the power rule, see the technical rules in Subpart E of Part 95.

CB Rule 21   May I use power amplifiers?

(a) You must not use or attach a linear or external radio frequency (RF) power amplifier at any CB station in any way.

(b) There are no exceptions to this rule.

(c) The FCC will presume you have used a linear or other external RF power amplifier if—

(1) It is in your possession or on your premises; and

(2) There is other evidence that you have operated your CB station with more power than allowed by CB Rule 20.

(d) Paragraph (c) of this rule does not apply if you hold a license in another radio service which allows you to operate an external RF power amplifier.

CB Rule 22   What communications may I transmit?

(a) You may transmit two-way plain language communications only to other CB stations, to units of your own CB station or to authorized government stations on CB frequencies about—

(1) Your personal or business activities or those of members of your immediate family living in your household;

(2) Emergencies (see CB Rule 25);

(3) Traveler assistance (see CB Rule 25); and

(4) Civil defense activities in connection with official tests or drills conducted by, or actual emergencies announced by, the civil defense agency with authority over the area in which your station is located.

(b) You may transmit a tone signal only when the signal is used to make contact or continue communications. (Examples of circuits using these signals are tone operated squelch and selective calling circuits.) If your signal is an audible tone, it must last no longer than 15 seconds at one time. If your signal is a subaudible tone, it may be transmitted continuously only as long as you are talking.

(c) You may transmit one-way communications for the purpose of voice paging.

(d) You may transmit in a foreign language, as long as you identify your CB station in the English language.

CB Rule 23   What communications are prohibited?

(a) You must not use a CB station—

(1) In connection with any activity which is against federal, state or local law;

(2) To transmit obscene, indecent or profane words, language or meaning;

(3) To interfere intentionally with the communications of another CB station;

(4) To transmit one-way communications, except for emergency communications, traveler assistance, brief tests (radio checks), or voice paging;

(5) To advertise or solicit the sale of any goods or services;

(6) To transmit music, whistling, sound effects or any material to amuse or entertain;

(7) To transmit any sound effect solely to attract attention;

(8) To transmit the word "MAYDAY" or any other international distress signal, except when your station is located in a ship, aircraft or other vehicle which is threatened by grave and imminent danger and you are requesting immediate assistance;

(9) To communicate with, or attempt to communicate with, any CB station more than 250 kilometers (155.3 miles) away;

(10) To advertise a political candidate or political campaign; (You may use your CB radio for the business or organizational aspects of a campaign, if you follow all other applicable rules.);

(11) To communicate with unlicensed stations or stations in other countries; and

(12) To transmit a false or deceptive communication.

(b) You must not use a CB station to transmit communications for live or delayed rebroadcast on a radio or television broadcast station. You may use your CB station to gather news items or to prepare programs.

(c) A CB station licensed to a telephone answering service must not be used to transmit messages to its customers. (See CB Rule 26.)

CB Rule 24   May I be paid to use my CB station?

(a) You must not accept direct or indirect payment for transmitting or receiving messages with a CB station.

(b) You may use a CB station to help you provide a service, and be paid for that service, as long as you are paid only for the service and not for the actual use of the CB station.

CB Rule 25   How do I use my CB station in an emergency or to assist a traveler?

(a) YOU MUST, AT ALL TIMES AND ON ALL CHANNELS, GIVE PRIORITY TO EMERGENCY COMMUNICATIONS.

(b) When you are directly participating in emergency communications, you do not have to comply with the rules about authorized users (CB Rule 26), length of transmissions (CB Rule 29), and communications with unlicensed stations (CB Rule 23). You must obey all other rules.

(c) You may use your CB station for communications necessary to assist a traveler to reach a destination or to receive necessary services. When you are using your CB station to assist a traveler, you do not have to obey the rule about length of transmissions (CB Rule 29). You must obey all other rules.

CB Rule 26   Who may operate under my license?

(a) You may permit only the persons listed below to operate under your license:

IF YOU ARE:	THE AUTHORIZED USERS ARE:
INDIVIDUAL	YOURSELF. Members of your immediate FAMILY living in your household. Each of your EMPLOYEES as long as his or her communications are *only* about your business.
PARTNERSHIP	Each PARTNER and EMPLOYEE of the partnership, as long as his or her communications are *only* about the business of the partnership.
ASSOCIATION	Each MEMBER of the association as long as his or her communications are *only* about the business of the association. Each EMPLOYEE of the association, as long as his or her communications are *only* about the business of the association.
CORPORATION	Each OFFICER, DIRECTOR and EMPLOYEE of the corporation, as long as his or her communications are *only* about the business or the corporation.
GOVERNMENTAL UNIT	Each EMPLOYEE of the governmental unit, as long as his or her communications are *only* about the business of that governmental unit.

(b) Someone else may operate your CB station if you request, and the FCC grants, special authorization to allow operation under your license where he or she would not otherwise qualify to operate your CB station.

(c) If you are a corporation, you may, upon request and FCC approval, permit your parent corporation or subsidiary corporation to provide you with a private radio-communications service under your license if the subsidiary or parent corporation provides the service on a non-profit or cost-sharing basis.

(d) (1) You may employ a telephone answering service to relay telephone messages to you on your CB transmitter if—

(i) You install or have someone else install a transmitter of your CB station at the answering service;

(ii) Your transmitter is used *only* under the authority of your license; and

(iii) Your transmitter is used *only* to relay messages to you about your personal or business affairs.

(2) If your transmitter is installed at a telephone answering service, it must not be used under the authority of any CB license other than yours.

(e) If you authorize any of the persons listed in paragraphs (a), (b), (c), or (d) of this rule to operate under your license, you must keep a list of all authorized users as part of your station records.

CB Rule 27   Who is responsible for transmissions made under the authority of my license?

You are responsible for all transmissions which are made by you or others under the authority of your license, including transmissions which are against these rules. Because you are responsible for all transmissions, you should be certain that anyone operating under your license understands and obeys the rules.

CB Rule 28   Who must not operate under my license?

(a) You must not permit anyone to operate under your license who is not listed in CB Rule 26, except in an emergency.

(b) You must not permit anyone who no longer has a CB license to operate under your license if—

(1) His or her license was revoked by the FCC; or

(2) His or her license was surrendered for cancellation after a notice of apparent liability to forfeiture was served by the FCC; or

(3) His or her license was surrendered for cancellation after the FCC instituted revocation proceedings.

(c) You must not permit anyone to operate your CB station if the FCC has issued a cease and desist order to that person, and the order is still in effect.

(d) You must not permit anyone to operate under your license if that person's most recent CB license application was denied by the Commission or dismissed with prejudice.

(e) If you sell CB transmitters, you must not allow a customer to operate a CB transmitter under the authority of your license.

CB Rule 29   Do I have to limit the length of my communications?

(a) Your communications must be limited to the minimum practical time.

(b) If you are communicating with another CB station or stations, you, and the stations communicating with you, must limit each of your conversations to no more than five continuous minutes.

(c) At the end of your conversation, you, and the stations communicating with you, must not transmit again for at least one minute.

CB Rule 30   How do I identify my CB communications?

(a) You must identify your CB communications by your FCC-assigned call sign at the end of each communication.

(b) Your FCC-assigned call sign must be clearly given in the English language. A phonetic alphabet may be used as an aid for identification. A "handle," unit designator, or special identifier may be used *in addition to,* but not instead of, your FCC-assigned call sign.

CB Rule 31   Where may I operate my CB station?

(a) You may operate your CB station in any of the fifty United States, in the District of Columbia, in Puerto Rico, in the United States Virgin Islands, on Guam, and in all other United States territories and possessions.

(b) You may operate your CB station in or on any aircraft or vessel of United States registry, with the permission of the appropriate officer.

(c) If your CB station is outside the fifty United States, the District of Columbia, Puerto Rico, the United States Virgin Islands, Guam or any of the other United States territories and possessions, you are subject to any applicable laws or regulations governing the location at which you are operating.

(d) You may operate your CB station in Canada, if you request and receive written permission *in advance* from the Canadian Department of Communications.

(e) If your CB station is located on land controlled by the Department of Defense, you may be required to comply with additional regulations imposed by the commanding officer of the installation.

CB Rule 32   May I operate my CB transmitter by remote control?

(a) You must not operate a CB transmitter by remote control, except as provided in paragraph (b).

(b) If you can show satisfactory need, the FCC may grant you written permission to operate by wire-line remote control. You must keep this permission as part of your station records. You can send your request for permission to the Personal Radio Division, FCC, Washington, D.C. 20554.

OTHER THINGS YOU NEED TO KNOW

CB Rule 33   How long must I keep my license?

You must keep your license (or other authorization) until it expires or until it is terminated.

CB Rule 34   Where must I keep my license?

(a) You must keep your license (or other authorization) in your station records or post it at your station.

(b) You may photocopy your license for any lawful purpose.

CB Rule 35   What do I do if I lose my license?

If you lose your license, you must request a duplicate license from the FCC, Gettysburg, Pa. 17326. Your request must include your name, your address and your station call sign.

CB Rule 36   Do I need to have a copy of the CB Rules?

(a) You must keep a current copy of the CB Rules in your station records. The CB Rules are published periodically by the Government Printing Office.

(b) You must stay up to date with changes to the CB Rules. Changes to the CB Rules are found in the *Federal Register* and in other publications.

(c) Your CB station must comply with technical rules found in Subpart E of Part 95, but you do not have to keep those rules in your station records.

CB Rule 37   What are the penalties for violating these rules?

(a) If the FCC finds that you have willfully or repeatedly violated the Communications Act, FCC Rules or 18 U.S.C. 1464 (which prohibits the transmission of obscene, indecent or profane language), you may have to pay as much as $2000 for each violation, up to a total of $5000. (See Section 503(b) of the Communications Act.)

(b) If the FCC finds that you have willfully or repeatedly violated the Communications Act or FCC rules, it may revoke your CB license. (Other grounds for revoking a CB license are listed in Section 312 (a) of the Communications Act.)

(c) If the FCC finds that you have violated any section of the Communications Act, you may be ordered to stop whatever action caused the violation. (See Section 312 (b) of the Communications Act.)

(d) If a federal court finds that you have willfully and knowingly violated any FCC rule, you may be fined up to $500 for each day you committed the violation. (See Section 502 of the Communications Act.)

(e) If a federal court finds that you have willfully and knowingly violated any provision of the Communications Act, you may be fined up to $10,000, or you may be imprisoned for one year, or both. (See Section 501 of the Communications Act.)

CB Rule 38   How do I answer violation notices?

(a) If it appears to the FCC that you have violated the Communications Act or these rules, the FCC may send you a written notice of the apparent violation.

(b) Within the time period stated in the notice, you must provide—

(1) A complete written statement about the apparent violation;

(2) A complete written statement about any action you have taken to correct the apparent violation and to prevent it from happening again; and

(3) The name and station call sign of the person operating at the time of the apparent violation.

(c) You must not shorten your response by references to other communications or notices.

(d) You must send your response to the office of the FCC which sent you the notice.

(e) If you cannot answer a violation notice within the time stated in the notice, because of illness or other unavoidable circumstances, you must answer at the earliest possible time and explain the reason for your delay.

(f) If the violation notice covers a violation related to technical transmitter standards, you must stop transmitting immediately, except for necessary tests and adjustments; and you must not transmit again until all technical problems with the transmitter have been corrected. The FCC may require you to have tests conducted and to report the results of those tests. (See CB Rule 41 for the rules about tests and adjustments.) Test results must be signed by the first or second class commercial radiotelephone operator who conducted or supervised the test or adjustment.

(g) You must keep a copy of your response as a part of your station records.

CB Rule 39   What must I do if the FCC tells me that my CB station is causing interference?

(a) If the FCC tells you that your CB station is causing interference for technical reasons, you must follow all instructions in the official FCC notice.

(b) You must comply with any restricted hours of CB station operation which may be included in the official FCC notice.

CB Rule 40   May I connect my CB transmitter to a telephone?

(a) You may connect your CB transmitter to a telephone if you comply with *all* of the following:

(1) You, or someone authorized to operate under your license, must be present at your CB station and must—

(i) Manually make the connection (the connection must not be made by remote control);

(ii) Supervise the operation of the transmitter during the connection;

(iii) Listen to each communication during the connection; and

(iv) Stop all communications if there are operations in violation of these rules.

(2) Each communication during the telephone connection must comply with all of these rules.

(3) You must obey any restriction that the telephone company places on the connection of a CB transmitter to a telephone.

(b) The CB transmitter you connect to a telephone must not be shared with any other CB station.

(c) If you connect your CB transmitter to a telephone, you must use a phone patch device which has been registered with the FCC.

CB Rule 41   How do I have my CB transmitter serviced?

(a) You may adjust your own antenna to your CB transmitter and you may make "radio checks."

(b) Each internal repair and each internal adjustment to your CB transmitter must be made by, or under the direct supervision of, a person holding a first- or second-class commercial radiotelephone operator license.

(c) Except as provided in paragraph (d) of this section, each internal repair and each internal adjustment of a CB transmitter in which signals are transmitted must be made using a nonradiating ("dummy") antenna.

(d) Brief test signals using a radiating antenna may be sent to adjust a transmitter to an antenna or to detect or measure spurious radiation. These test signals may not be longer than one minute during any five minute period.

CB Rule 42   May I make any changes to my CB transmitter?

(a) You must not make or have anyone else make *any* internal modification to your CB transmitter.

(b) You must not operate a CB transmitter which has been modified by *anyone* in any way, including modification to operate on unauthorized frequencies or with illegal power.

CB Rule 43   Do I have to make my CB station available for inspection?

If an authorized FCC representative requests to inspect your CB station, you must make your CB station available for inspection.

CB Rule 44   What are my station records?

(a) Your station records include the following documents, as applicable:

(1) Your temporary permit (CB Rule 6);

(2) A copy of each letter telling the FCC of your name or address change (CB Rule 14);

• (3) Your license (CB Rule 34);

(4) A list of authorized users of your CB station (CB Rule 26);

(5) A current copy of the CB Rules (CB Rule 36);

(6) A copy of each response to an FCC violation notice (CB Rule 38);

(7) Each written permission received from the FCC.

(b) If an authorized FCC representative requests to inspect your station records, you must make your station records available for inspection.

(c) You must keep your station records for the term of your license.

CB Rule 45   How do I contact the FCC?

(a) You may write to the following address about your application, about the rules or when you are requesting permission to use more than 25 transmitters:

> Personal Radio Division, FCC, Washington, D.C. 20554.

(b) You may write to the following address when you send your notice of new name or address, or when you send a new or renewal application form:

> FCC, Gettysburg, Pa. 17326.

(c) You may write to any of the following FCC offices in the field if you wish to file an interference complaint. The FCC will forward your complaint to the appropriate field enforcement unit.

Alaska, Anchorage 99510, FCC, room G-63, U.S.P.O. and Courthouse Bldg., P.O. Box 644, 4th and F Sts.

California, Long Beach 90807, FCC, room 501, 3711 Long Beach Blvd.

California, San Diego 92101, FCC, Fox Theatre Bldg., 1245 7th Ave.

California, San Francisco 94111, FCC, 323-A Customhouse, 555 Battery St.

Colorado, Denver 80202, FCC, suite 2925, The Executive Tower, 1405 Curtis St.

Florida, Miami 33130, FCC, room 919, 51 Southwest 1st Ave.

Florida, Tampa 33602, FCC, Barnett Office Bldg., room 809, 1000 Ashley Dr.

District of Columbia, Washington 20554, FCC, 1919 M. St. N.W., room 411.

Georgia, Atlanta 30309, FCC, room 440, Massell Bldg., 1365 Peachtree St. N.E.

Hawaii, Honolulu 96808, FCC, 502 Federal Bldg., P.O. Box 1021, 355 Merchant St.

Illinois, Chicago 60604, FCC, 230 South Dearborn St., room 3935.

Louisiana, New Orleans 70130, FCC, 829 F. Edward Hebert Federal Bldg., 600 South St.

Maryland, Baltimore 21201, FCC, 819 Federal Bldg., 31 Hopkins Plaza.

Massachusetts, Boston 02109, FCC, 1600 Customhouse, 165 State St.

Michigan, Detroit 48226, FCC, 1054 Federal Bldg., 231 West LaFayette St.

Minnesota, St. Paul 55101, FCC, 691 Federal Bldg. and U.S. Courthouse, 316 North Robert St.

Missouri, Kansas City 64106, FCC, 1703 Federal Bldg., 601 East 12th St.

New York, Buffalo 14202, FCC, 1307 Federal Bldg., 111 West Huron St.

New York, New York 10014, FCC, 201 Varick St.

Oregon, Portland 97204, FCC, 1782 Federal Office Bldg., 1220 Southwest 3rd Ave.

Pennsylvania, Philadelphia 19106, FCC, James A. Byrne Federal Courthouse, 601 Market St.

Puerto Rico, Hato Rey 00918, FCC, room 747, Federal Bldg.

Texas, Dallas 75242, FCC, Earle Cabell Federal Bldg., U.S. Courthouse room 13E7, 1100 Commerce St.

Texas, Houston 77002, FCC, New Federal Office Bldg., 515 Rusk Ave., room 5636.

Virginia, Norfolk 23502, FCC, Military Circle, 870 North Military Highway.

Washington, Seattle 98174, FCC, 3256 Federal Bldg., 915 2nd. Ave.

CB Rule 46   How are the key words in these rules defined?

In the CB radio rules, the following definitions apply:

Antenna structure means the antenna's radiating system, the antenna's supporting structure, and anything mounted on the antenna or its supporting structure.

Carrier power means the average power at the output terminals of a transmitter (other than a single sideband unit or a transmitter with a suppressed, reduced or controlled carrier) during one radio frequency cycle under conditions of no modulation.

CB station means a station licensed in the Citizens Band (CB) Radio Service. It includes all of the radio equipment you use.

Emergency communications mean messages concerning the immediate safety of life or the immediate protection of property.

External radio frequency power amplifier means any device which is not included by the manufacturer in a type-accepted transmitter and which, when used with a radio transmitter as a signal source, is capable of amplifying that signal. (External radio frequency power amplifiers are sometimes known as "linears".)

Mailing address means the place where you receive your mail.

One-way communications means a message which is not intended to establish communications with one or more particular CB stations.

Peak envelope power (used by SSB units) means the average power at the output terminals of a transmitter during one radio frequency cycle at the highest crest of the modulation envelope, taken under conditions of normal (voice) operation.

Person means an individual, a partnership, an association, a joint-stock company, a trust or a corporation.

Plain language communications means communications without codes or coded messages. (Operating signals such as "ten codes" are not considered codes or coded messages.)

Remote control means operation of a CB transmitter from any place other than the location of the transmitter. Direct mechanical control or direct electrical control by wire from some point on the same premises, craft or vehicle as the transmitter is not considered remote control.

Single sideband emission means an emission in which only one sideband is transmitted. The carrier, or a portion of the carrier, may be present in the emission.

Double sideband emission means an emission in which both upper and lower sidebands are transmitted. The carrier, or a portion of the carrier, may also be present in the emission.

Station address means the place where the station license is kept or posted (see CB Rule 34), where the station records are kept (see CB Rules 34 and 44) and where the primary fixed transmitter (if any) is operated.

Station authorization means a CB temporary permit or a CB license or special temporary authority issued by the FCC.

Subaudible tone means any tone or combination of tones having only frequencies below 150 Hertz.

Voice paging means directing a message to a particular CB receiver (or receivers) solely for the purpose of transmitting a particular communication to that receiver (or receivers).

Index

A

Advisory, frequency, 44
Aeronautical radiocommunication
    and radionavigation, 39-41
Aircraft radio installation, 40
Antenna
    ammeter, 76
    monitor, 76
Assignments, paired frequency, 69
Aviation Radio Services, 35-44

B

Broadcast of
    recorded material, 107
    telephone conversations, 108

C

Categories, station, 37-39, 52-59,
    61-62
CB base station, 64
Centers, radio control, 44-48
Channel designations, 31
Citizens Band Radio Service, 63-68
Classifications of airborne stations,
    37
Coast Guard, 28
Communications
    identification of, 19
    nature of, 18
    priority of, 18-19
Considerations, license, 8-17
Control, remote, 90-95

D

Definitions, 104-105
Direction finding receiver, 36
Distress procedure, 21-22
    radiotelephone, 22-23
Duplicate and renewed commercial
    operator permits, 109-110

E

Electrical terms, 105
Element
    I questions, 128-129
    I and II self test, 139-144
    II, 130-138
Emergency and distress, 41-42
Emergency Broadcast System,
    108-109

Equipment tests, 42
Essential provisions for radio
    operators, 17-22

F

FCC field offices, 12-13
Final-plate
    -current meter, 75
    -voltage meter, 75
Frequencies and range, 27
Frequency advisory, 44

G

General Mobile Radio Service,
    68-71

H

Hand-held
    CB transceiver, 65
    marine transceiver, 34

I

Identification
    of communications, 19
    sponsor, 107 ·
    station, 42-43, 107
Important rules, 27-28
IMTS
    mobile telephone, 57
    telephone system, 59-61
Industrial Radio Services, 48-61
Interference, prevention of, 19

K

Kilohertz, 2182, 29

L

Land Transportation Radio
    Services, 61-62
License
    considerations, 8-17
    operator, 8
    station, 8, 17
Lighting, tower, 109
Log
    operating, 98-100
    requirements, 98-100
Lotteries, 108

M

Malfunctions, transmitter, 100-102

Marine radiotelephones, 30-34
Maritime Radio Services, 24-26
Megahertz, 156.8, 30
Meter
   final-plate-current, 75
   final-plate-voltage, 75
   modulation, 75
   power-output, 75
Metering, transmitter, 75-77
Mobile CB transceiver, 65
Modulation meter, 75
Monitor, antenna, 76
Monitoring summary, 87-90
Motor carrier radio, 62

N

Nature of communications, 18
Notice of violations, 20

O

156.8 megahertz, 30
Operating
   log, 98-100
   power, 77-79
   procedure, 28
Operation, station, 42
Operator
   license, 8
   licenses or permits, 17-18
   permits, posting of, 106
   requirements, 26, 37, 43-44, 51-52
   responsibilities, 74-75

P

Paired frequency assignments, 69
Penalties, 20-21
Plans, station, 79-81
Posting of operator permits, 106
Power
   operating, 77-79
   -output meter, 75
Prevention of interference, 19
Priority of communications, 18-19
Privileges, special, 15-17
Procedure, distress, 21-22
Public Safety Radio Services, 43-48

Q

Questions on Element I, 128-129

R

Radio
   -control centers, 44-48
   log, 19-20, 28-29
   motor carrier, 62
   Services, Aviation, 35-44

Radiocommunications, secrecy of, 19
Radiotelephone
   distress procedure, 22-23
   marine, 30-34
Railroad, radio, 62
Rebroadcast, 107
Recorded material, broadcast of, 107
Reference tower, 77
Remote control, 90-95
   equipment and operation, 105-106
Requirements
   log, 98-100
   operator, 26, 37, 43-44, 51-52
Responsibilities, operator, 74-75

S

Secrecy of radiocommunications, 19
Self-test for Elements I and II, 139-144
Simplex operation, 50-51
Special privileges, 15-17
Sponsor identification, 107
Station
   categories, 37-39, 52-59, 61-62
   identification, 42-43, 107
   inspection and availability of records, 110
   license, 8, 17
   operation, 42
   plans, 79-81
   requirements and operator duties, 95-98
Synthesized scanning marine radiotelephone, 33

T

Technical considerations for radio broadcasting, 73-74
Telephone conversations, broadcast of, 108
Terms, electrical, 105
Tests, equipment, 42
Third-class operator permit, 8
Tower lighting, 109
Transmitters, 29
   malfunctions, 100-102
   metering, 75-77
2182 kilohertz, 29

V

Violations, notice of, 20
VOR or OMNI stations, 40